INFORMATION POLICY:
A FRAMEWORK FOR EVALUATION
AND POLICY RESEARCH

Information Management, Policy, and Services
Charles R. McClure and Peter Hernon, Editors

INFORMATION POLICY:
A FRAMEWORK FOR EVALUATION
AND POLICY RESEARCH

by
ROBERT H. BURGER
University of Illinois at Urbana-Champaign

ABLEX PUBLISHING CORPORATION
NORWOOD, NEW JERSEY

The author wishes to thank Harper Collins Publishers for permission to quote from Frederick Buechner's work *Now and Then* (1983, p. 83) for the dedication and to MIT Press for permission to reproduce the chart from Arnold Pacey's *The Culture of Technology* (copyright 1983, MIT Press, p. 6) for Chapter 9, Figure 9-1

Library of Congress Cataloging-in-Publication Data

Burger, Robert H. (Robert Harold), 1947–
    Information policy : a framework for evaluation and policy research / by Robert H. Burger.
       p. cm. — (Information management, policy, and services)
    Includes bibliographical references amd index
    ISBN 0-89391-890-3 (cl). — ISBN 1-56750-018-8 (ppb)
    1. Information services and state. 2. Information services and state—United States. 3. Communication in science—Government policy. 4. Communication in science—Government policy—United States. 5. Communication of technical information—Government policy. 6. Communication of technical information—Goverment policy—United States. 7. Science and state. 8. Science and state—United States. 9. Technology and state. 10. Technology and state—United States. I. Title. II. Series.
Z674.4.B87 1992
020'9733—dc20
                                     92-42683
                                       CIP

Ablex Publishing Corporation
355 Chestnut St.
Norwood, NJ 07648

This book is dedicated to
RUTH R. HANSEL

Happy the man whose mother-in-law turns out to be, not the comic-strip adversary, but as dear and valued a friend as any he has.

Frederick Buechner

CONTENTS

LIST OF FIGURES

LIST OF TABLES

ACKNOWLEDGMENTS

I would like to acknowledge with much gratitude the following persons who gave me both direct and indirect help with this book. Peter Hernon and Charles McClure were patient and helpful editors; Leslie Edmonds, Leigh Estabrook, Bruce Hannon, F.W. Lancaster, and Linda Smith were encouraging and demanding during the dissertation phase of this study (chapters 6–8); Clifford Christians introduced me to the writings of Arnold Pacey and Manfred Stanley and taught me that technology is not value-free; Pat Stenstrom, Sandy Wolf, and Melissa Ritter of the University of Illinois Library and Information Science Library were models of information service providers; all my colleagues at the Slavic and East European Library, especially Larry Miller, Dmytro Shtohryn and Helen Sullivan made possible the freedom from everyday concerns during my sabbatical; my daughters, Sarah, Meg, and Nell have continued to keep me aware of the existence of life outside of academia; and my wife Ann has endured with grace and good humor more conversations about information policy and its implications than anyone deserves. The book is dedicated, however, to Ruth Hansel, my favorite mother-in-law, without whom my life would have been poorer by lacking one good friend and counselor.

PREFACE

This book is about the use of specialized knowledge in evaluating and designing information policy. Such a statement implies that such specialized knowledge already exists and the material presented here is simply a discourse on applying such knowledge to produce effective information policy. Such an inference on the reader's part, however, would be unwarranted. Indeed, part of the purpose of this book is to discover what that specialized knowledge actually is and what it should be.

The origin of this book dates back to 1982 when I made the unremarkable discovery that librarians and information scientists played a relatively minor role in preparing many of the information policy proposals and studies issued over the past thirty years. I refer, of course, to the *Weinberg Report*, the *SATCOM Report*, the *Greenberg Report*, *The Privacy Study Commission Report*, *The Attorney General's Report on Pornography*, the *Rockefeller Commission Report*, and *Scientific Communication and National Security*, among others. I speculated at the time that there were many possible reasons for this underrepresentation of information professionals in the policy-making process. For example, the information specialists may not have possessed the power and influence that is often needed to participate in the process; they may have not been recognized as possessing the skills and knowledge necessary to formulate policy proposals; or worse yet, there may not have been any specialized corpus of knowledge in the information field upon which to draw for such policy making. Combinations of these and other explanations were also possible.

In examining the situation further, I became more interested in my supposition that the existing corpus of knowledge about information processes, at least that corpus taught in schools of library and information science and information studies in the United States, focuses for the most part on the behavior of specific, institution-bound information systems and their users. Might it be, I wondered, that our existing knowledge, robust as it was for microlevel investigations, was inadequate for the task of examining societal information flows and information policy? Lancaster and Burger (1990) explored this suspicion in a recent paper.

We reasoned that in order to evaluate information policy adequately, we need to develop two related umbrella disciplines called, for want of better names, macroinformatics and microinformatics. These "disciplines" are currently only figments of our imaginations. They have been defined in the following way: "We propose the term macroinformatics to refer to the study of information transfer phenomena 'in the large,' without reference to the views or interests of particular segments or elements of society, while microinformatics would imply the study of information transfer phenomena from the varying perspectives of these particular segments or elements" (Lancaster and Burger, 1990, p. 153). We further argued for the importance of this distinction and its applicability to all information policies, irrespective of their type, their scale, or the countries in which they are promulgated, just as the distinction between macroeconomics and microeconomics or macrosociology and microsociology are important for economics and sociology and the phenomena that are studied by these disciplines.

If macroinformatics would exist, then it might be used in the policy-making process (the making of political choices that result in the production of government actions). Once a policy produces its effect on society, then this knowledge could also be used to evaluate the effectiveness and wisdom of the policy.

I decided to pursue this idea further and began a limited quest in search of macroinformatic knowledge in the discipline of information science. Unfortunately, I was unable to find any consensus in the literature concerning the domain of information science and the theoretical constructs germane to it. This was not surprising in and of itself, but interdisciplinary squabbles aside, I discovered that domain descriptions that were the most specific did not offer the macrolevel information I sought; those that cast their net wider in identifying the domain were diffuse and simply included other disciplines, such as sociology, economics, and management, into the domain. Although macroinformatics could be an interdisciplinary conglomeration of several fields, there did not seem to be any comprehensive macrolevel theory or empirical observations in information science or fields often related to it that could be readily applied in the evaluation of information policy.

If we did not now have a coherent body of knowledge that we might use in comprehensively evaluating information policy, then how far did our current knowledge actually take us toward this goal? To explore this idea further, I conducted a case study (the essence of which appears in Chapters 6 through 8) in order to determine the consensus of information professionals about information science's applicability to solutions proposed in the SATCOM Report (*Scientific and Technical Information*, 1969). The results suggested, among other things, that, in fact, there was consensus about the applicability of information science to problems on a microlevel, but less consensus on the applicability of information science to problems on a macrolevel.

This case study led me to broaden the context for examining the role of specialized knowledge in the evaluation of information policy. This book is the result of that examination. It consists of four parts. Part I, Chapters 1–3, examines various contexts in which information policy is made. Chapter 1 describes the context in which American information policy is made. Whenever possible, I have endeavored to use examples of existing information policies to inform the discussion of the general policy making process. This examination is expanded in Chapter 2 to illustrate examples of international information policy. The purpose in this chapter is to suggest that problems of dealing with the making of information policy are similar abroad, in spite of the different political and societal cultures in which policy is made. Furthermore, it also emphasizes the complexity of information policies that are international in scope or that have international implications.

Chapter 3 narrows the discussion to scientific and technical information policy. Here its characteristics and importance are described. I speculate that in the near future specialized knowledge will probably be most useful in the making of this type of information policy. The material presented here will suggest that we are not doing a very good job of creating such policies. These three chapters set the groundwork for Part II (Chapters 4 and 5).

Chapter 4 discusses the desiderata of information policy making, while Chapter 5 explains why these desiderata are difficult to achieve in light of a review of the existing literature of information science and information policy. One way of overcoming the difficulty is by bringing specialized knowledge to bear. Does information science now offer us this usable knowledge, or are other approaches necessary? Chapter 5 attempts to discern the possibilities for this type of application by describing what others have written about the domain of information science. Such knowledge would ideally allow us to define and resolve the problems and complexities we face in information policy making.

A case study of the applicability of this type of specialized knowledge to the SATCOM Report is contained in Part III (Chapters 6–8) by covering the method of investigation, an examination of a randomly selected number of policy choices, and a summary of this examination. Part IV (Chapter 9) suggests a new, broadly conceived research agenda within the framework of the evaluative method presented here and describes an expanded view of the specialized knowledge required to carry out that agenda.

I believe we face a crisis in information policy making. The crisis is not that we are hampered from making information policies. Very little seems to dissuade us from doing that. The crisis is in our ability to evaluate and design these policies in the best ways possible. We have been doing evaluation of information policies by using several disciplines available to us: law, economics, political science, and ethics. It seems to me, however, that although our intent has been noble and our evaluations extensive, we have missed the central point of evaluation. Evaluation, I believe, should deal with the deceptively simple questions: Will this policy have the effect we intend for it to have? Will the policy have effects that we would rather not think about? Will (and perhaps this is the central question) the disciplines we are using to analyze this policy give us these answers?

This, of course, raises a whole host of questions that this book is not intended to solve, but are logical outcomes of this preliminary examination. For example, is there a predictive social science related to information policy that can tell us whether a specific policy will or will not work? Is there a basic organizational framework for constructing information models equivalent to supply-and-demand models in economics or conflict models in sociology? What are the problems confronted by information policies: the allocation of scarce information resources, the efficient utilization of information resources, maximizing information use, or some principle that is operative but not yet sufficiently articulated (Lancaster and Burger, 1990, pp. 155–156)? This book will begin to examine information policies and the relation of information science to it. In doing so, I would like to encourage further research so that we may begin the more difficult and extensive task of answering the questions posed above.

PART I

THE CONTEXTS
OF INFORMATION
POLICY MAKING

CHAPTER 1

INFORMATION POLICY IN THE PUBLIC POLICY ARENA

Information policy is one of many different types of public policies that governments make. Among public policies, however, it is a relative infant. In spite of the fact that scientific and technical information policies have been suggested since at least the early 1960s (see, e.g., Weinberg, 1963), a comprehensive bibliography of policy studies literature (Nagel, 1984) did not include information policy among its specific policy problems section, nor was any substantial part of the section on Communication Policy devoted to it. This latter section focused mainly on broadcasting, telecommunications, spectrum allocation, cable systems, and transnational data flows. These topics certainly fall within the broad realm of information policy, but information policy also includes much more, such as scientific and technical information policy, privacy issues, literacy, freedom of speech, libraries and archives, secrecy and its effects on commercial information policy and national security, and access to government information. Indeed, the literature of the time (early 1980s) had not yet formed into an identifiable field within the broad realm of policy analysis. There had been discussions of these issues, to be sure, but no thread sewed these various policy fragments together into a recognizable garment.

Over the past decade, however, this situation has changed in some respects. Information policy courses have emerged in some graduate schools of library and information science; several journals that focus directly or tangentially on the topic are being published; and a monographic series, of which this volume is part, that deals directly with information resources management is now being issued. In a very real sense, information policy has come of some age as a subject for study. Diffusion, however, is still a hallmark of the field.

3

IMPORTANCE OF INFORMATION POLICY

According to some, information policy affects every aspect of our lives; our economic well-being is dependent on it; our constitution is alternately assailed or defended as a result of it. We are told, over and over again, often in cliched terms, about the importance of information policy in our "information society." The issues are inevitably laid before us either with an almost morbid sense of doom—there are tradeoffs between freedom of information and economic exigencies, between corporate rights and private rights, between national security and open communication in science, between freedom of speech and moral decay—or with an unbounded optimism that portrays the future as a paradise of human, economic and cultural freedom and opportunity. Dizard (1982, p. 2) has warned us about the gravity of the decisions we must make:

> One characteristic of the new age stands out among the welter of trends. This is the increasing emphasis on the production, storage, and distribution of information as its major activity. Our strategy for organizing the transition to this new environment will set the quality and pattern of American life well into the new century.

Along with the descriptions of our new information age come analyses of every imaginable kind: legal, economic, classificatory, systemic, political, procedural, and providential. These analyses are used on both sides of the information debate to support conflicting arguments.

Further, although information policy is deemed important, it is only brought to the public's attention through dramatic cases such as the incident of the Robert Maplethorpe exhibit in Cincinnati and the attendant questions of Federal support for the arts, the burning of the flag issue, the 1973 law on obscenity and the arrest of retailers for selling 2 Live Crew tapes, or the prohibition of telecommunication firms from setting up a modernized telecommunications system in the Soviet Union. The issues arise frequently and with much political passion. Any informed citizen cannot escape knowledge of these problems.

We often forget that the main issue is not the specific case before us, however important it may seem at the time, but generally how information of any kind is used and who either has the power or is allowed to control the information. These global issues of control over information are political choices that each society must make. The tools for controlling information flows are information policies.

Yet, for the most part, we tend to think about these issues in reductionist terms. That is we use a limited number of analytical approaches in discussing them and resolving them. We not only do this as citizens but the policy analysts do this as well. For example, A. A. Bushkin and J. H. Yurow, the former a Special Assistant for Information Policy to the Assistant Secretary of Commerce for Communications and Information and the latter a Senior Policy Analyst at the National Telecommunications and Information Administration, in a paper discussing the foundations of United States information policy, boldly state that "aspects of United States domestic information policy can be divided into two broad categories: (1) the legal foundations of information dissemination and access; and (2) the economics and management of information" (Bushkin and Yurow, 1981, p. 217). Although this statement was written over ten years ago, many analysts still implicitly adhere to this analytical framework in formulating and evaluating information policy.

There are certainly understandable reasons for our difficulty in understanding information policies. First, information policies deal with information. In spite of several attempts to define information as a resource, to attribute quantifiable values to it, or to explain it in an epistemological hierarchy of data–information–knowledge–wisdom, it remains an intangible enigma. Even though we can create, analyze, synthesize, store, sell, and distribute it, among other things, we cannot always discuss it cogently and without some degree of ambivalence or ambiguity. Therefore, there is some merit to Bushkin's and Yurow's scheme because it reduces the ambiguity in discussing and analyzing the relevant policies.

Second, information policy deals with policy. Again, this is not a remarkable insight, unless one is aware of all the policy-related research that now exists. Political scientists themselves have a difficult time defining and understanding policy. Information policy is no different.

A third reason for information policy's elusiveness is the inescapable perception that it is involved with every social choice we make. Indeed, Mason (1983, p. 93) argues that

> in the United States, "information policy" is actually a set of interrelated laws and policies concerned with the creation, production, collection, management, distribution and retrieval of information. Their significance lies in the fact that they profoundly affect the manner in which an individual in a society, indeed a society itself, makes political, economic and social choices.

That is, in order to make any type of policy or choice, at any level, we humans require information. The ways we obtain information, even information from friends and relatives, are governed by extant information policies, policies formulated at all levels of society, from the individual to the nation state, either formally or informally. In this sense, information policy affects virtually everything we do, and every other type of policy that we mandate.

Fourth, it is complex because we have not yet developed ways of understanding it that are widely accepted and are broad enough to encompass its range. We now rely on existing disciplines to inform us about the soundness of information policies. But if information policy itself affects these disciplines, by controlling information flows to disciplinary practitioners, and these disciplines in turn affect information policy, then we may be running in circles by not looking beyond the disciplinary boundaries of, for example, law, political science, economics, and management for answers to information policy problems. Furthermore, we are relatively ignorant about the relationship of various types of information policies to one another. We only have a rudimentary topology of information policies based on active areas. Chartrand and others have classified information policy into the following categories: (1) federal information resources management; (2) information technology for education, innovation, and competitiveness; (3) telecommunications, broadcasting, and satellite transmissions; (4) international communications and information policy; (5) information disclosure, confidentiality, and the right of privacy; (6) computer regulation and crime; (7) intellectual property; (8) library and archives policies; and (9) government information systems clearinghouses and dissemination (Chartrand, 1986a; Milevski, 1986). Certain other topologies, for example, based on overall effect are certainly in order, but we have not yet progressed this far.

The current state of our knowledge about information policy, that is, the societal mechanisms used to control information, and the societal effects of applying these mechanisms demand that we examine information and information policy to see if we now possess the knowledge to carry out the task at hand. What is needed, I believe, is to find a legitimate disciplinary home for information policy analysis. This disciplinary home is necessary at the very least to coordinate our existing knowledge about the many aspects of information policy and its effects on society. It may very well be that we will be forced to produce new knowledge within the disciplinary framework in order to carry out our self-imposed charge. It may also turn out that this discipline will draw on techniques and approaches that already exist in other disciplines and will apply them to information policy.

The contention made here is that information science has the *potential* to be this disciplinary home. Information science is a relatively new discipline, characterized in one writer's words by "terminological chaos" that reflects a "chaos in conceptualization" (Schrader, 1984, p. 237). In general, practicing information scientists see their discipline "as primarily practical and technological" (Boyce and Kraft, 1985, p. 155), one that deals with problems in information theory, information retrieval and bibliometrics. It tends to focus on what has been called microinformatics issues (Lancaster and Burger, 1990) that deal with information transfer phenomena from the perspective of particular societal elements and institutions.

The discipline demanded by the problems posed by information policy, however, must include not only the micro perspective, but a macro perspective that concerns itself with "information transfer 'in the large,' without reference to the views or interests of particular segments or elements of society" (Lancaster and Burger, 1990, p. 153). Indeed, the definition of informatics put forth by the French Academy in 1966, as "the science concerned with semantic processing, especially by automatic machines, of information considered as the basis of human knowledge and communication in the technical, economic and social fields" (Gilyarevsky, 1990, p. 160, quoting Le Garff, 1982), would include this broader perspective. Gilyarevsky points out that the Russians have used this term

> to denote the field of knowledge studying the structure and general properties of scientific information, as well as the basic regularities of all processes of scientific communication . . . [i.e.] the sum total of processes involved in presenting, transferring and obtaining this information in science, technology, national economy and other social spheres. (Gilyarevsky, 1990, p. 160; Mikhailov, Chernyi, and Giliarevskii, 1984, Chapter 10)

In order to develop such a discipline, we need to take stock of the context in which information policy is made.

Since information policy is made within the context of politics, we must first turn, I believe, to the larger discourse on public policy and public policy analysis. If we do this we will be able to use many relevant findings and apply decades of thinking about public policy, its analysis and evaluation.

In order to bring us into this discourse, it would be helpful to sketch the policy making process and the role that policy analysis plays in it. By doing this, the evaluation method presented in Chapter 6 will make more sense and we can visualize the deficiencies that presently exist in the study of information policy. One risk taken in this endeavor is repeating, in too brief a fashion, information about which others have written volumes. My purpose is not, therefore, to break any new ground in this area, but specifically to describe the arena in which information policy functions and the ways in which others have attempted to analyze this arena.

THE POLICY PROCESS[1]

Public policy has been defined as "a set of interrelated decisions taken by a political actor or group of actors concerning the selection of goals and the means of achieving them within a specified situation where these decisions should, in principle, be within the power of these actors to achieve" (Jenkins

---

[1]Part of what follows here was presented in a slightly different form in Burger (1986).

1978, p. 15). This definition, however, says relatively little about how those decisions are made, by what authority they are made, the constraints put upon political actors in making those decisions, the design of the institutional setting in which political choices are made, and the agencies that produce the outcomes of these political decisions. All of these questions are germane if we wish to understand the policy process. Policy theorists have attempted to divide the process of making public policy into several stages. The three stages can be defined as policy formation or policy making, policy implementation or policy production, and policy evaluation and feedback (Kelman, 1987).

Policy Formation

Policy formation or policy making covers the policy process from the inception of an idea to its passage as a law by a legislative body. Ideas for public policies can and do come from anyone, both inside and outside the government. With information policy, for example, ideas can come from a local antipornography group, a cable television franchise owner, a vice-chancellor at a major research university, or a bureaucrat in the Office of Management and Budget. In order for policy making to take place, however, people who have the authority to make the policy decisions and those who have the resources and ability to influence those policy makers must be involved. These actors attempt to influence policy making to achieve their own ends.

Besides the traditional lobby organizations, such as the Information Industry Association, those attempting to influence policy may also include groups of private individuals specifically called together for the purpose of producing ideas that can then be forwarded to legislators for possible action. An example of this type of group is the 1991 White House Conference on Library and Information Services. A more structured and politically inflamatory group was the Attorney General's Commission on Pornography. Ideas from members (policy actors) of these groups may be brought together to help solve a particular problem, suggest possible solutions for legislators to consider, or simply to gather information about a problem confronting society. The authority of these actors is limited by the way in which they are permitted to interact with government representatives or by laws that only extend policy making powers to elected officials, such as a legislator or a member of a city council.

Policy actors from within the government can be at any level, have differing amounts of authority for policy making, and have various constraints on their activity. If they are within the bureaucratic hierarchy (as opposed to those who can, for all intents and purposes, ignore the bureaucratic constraints), they probably must follow organizational rules that limit their activity and influence. The director of the National Commission for Library and Information Science, for example, has no authority to implement suggestions gathered as a result of studies sponsored by the Commission. The director, however, by virtue of his position, may be able to persuade some members of Congress to vote for a specific law that will increase access to government information.

If the policy actors are elected officials, the ever-present concern of reelection may influence their behavior and, therefore, the scope and strength of their influence on certain issues are altered accordingly. Legislators tend to be more responsive to their constituents than to a more diffuse national public good. The range of behavior that they may exhibit and the power that they wield are directly tied to the institutional design of the government in question. In the United States, for example, the Commissioner of Copyrights cannot alter portions of the Copyright law enacted by Congress. The Commissioner has the responsibility to carry out the law as written.

"Thus policy formation usually involves a diverse set of authoritative, or formal, policy makers, who operate within the government arena, plus a diverse set of special interest and other constituency groups from outside arenas, who press their demands on these formal leaders" (Nakamura and Smallwood, 1980, p. 32). As discussed in the section on implementation, policy actors also play an important role in implementing a policy once it has been formed. These policy actors may include some of the same people that were active in formulating the policy.

Policy formation also produces policy goals and instructions. In order to carry out a policy effectively, or even to make a policy, the actors presumably have a goal in mind. This goal may be technical, social, or political; it may be implicit or explicit. The difficulty with analyzing this stage of policy making is identifying the goals implied by the policy. Some policies do not have explicitly stated goals. This, or course, will effect how the policy is implemented.

Stages in Policy Formation

The Proposal. Up to this point we have been thinking in general terms about the formation of a policy. There are, however, several distinct stages in formulating a policy. The first stage is the proposal. At this point a person or a group of persons has an idea about how to solve a problem or effect some change. They would propose an action to be taken and would also present some type of rationale for this proposed action. This rationale may or may not be based on good political sense, on existing scientific knowledge, or on all the facts available. But it is based on some rationale for action. Acknowledgement of the fact that a rationale exists does not necessarily mean that we must also adhere to the rational model of decision making.

Using this model, policy analysts have attempted to identify what takes place in policy making. In this model,

> A rational actor first clarifies the relevant values and corresponding goals, then lists all important alternative strategies for achieving these goals, and finally chooses the stategies [sic] judged to be optimal. To make this classical notion of rational action at the individual level applicable to macro-decision in public policy making, it must also be assumed that policy makers agree on policy goals. (Albaek, 1989–1990, p. 8)

As many writers on policy making have shown, political actors and decision makers rarely act this way. In fact, trying to come up with a model for how policy makers make policy has been a particularly active endeavor for policy analysts. Lindblom (1980) may be consulted for an intelligent discourse on this problem.

One principle, however, that anyone involved with public policy must acknowledge is that the proposed action is based on some rationale. At times this rationale may be made explicit in a proposed policy document; at other times the rationale may be kept secret for political or personal reasons and not discovered or divulged until after the policy has been implemented. The nonpersonal rationale, however, must be divulged to more than one person even at this stage of policy making. It must be divulged because this is the only way someone may persuade another that the proposed policy is both worthwhile, achieving some desired goal, and likely to achieve that goal. The value of the policy can be dependent on ideology or personal preference. The likelihood of achieving that goal, however, is open to scientific (to use the term loosely) scrutiny. It is at this point that scientific, social scientific, or political knowledge can enter the picture in order to ascertain whether the desired goal can be met.

The likelihood that a policy will reach the desired goal is, of course, dependent on the political skill of its backers. It is also dependent on theoretical knowledge that at some level predicts that if such and such an action is carried out, an expected effect will ensue. For example, placing a higher tax on gasoline, should, other things being equal, result in lower gas consumption because the "laws" of supply and demand predict that a higher price for an item will reduce demand for an item. Unfortunately, it is often difficult to discern the difference between speculation and sound, empirically-based theory.

Evaluation Stage of Policy Formation. Once the rationale for a policy is divulged to another, some type of evaluation of the rationale takes place. This evaluation may be informal and take the form of internal debate on the merits of the proposed action and the reasons for undertaking it. It may be formal, in the sense that policy analysts look at the policy in order to see the various implications of the proposed actions. The evaluation can, but does not always, take into consideration the various possible rationales for the action and the way the backer of the policy attempts to persuade other potential backers. The evaluation may at this point be cursory or in-depth. Relevant scientific, social scientific , or political knowledge may or may not be brought to bear. In established areas of policy making, such as economics and law, the relevant knowledge is common knowledge among the evaluators and the disciplines producing this knowledge are well established. The question that this book addresses, however, is what happens when a policy is proposed and the policy area is new, such as information policy, and the discipline producing knowledge about information, its creation, distribution, storage, and use is a nascent one? How can we approach such policy areas with a modicum of knowledge about the effects of this policy beyond the economic and legal effects? It is a fascinating question. One that at present, I fear, has no ready answer.

Once the policy is evaluated at this stage, prior to its being given a wider audience, it must then be placed on the agenda of the legislative or organizational body that is given the responsibility for enacting laws and regulations. Since this stage is of minor concern in this book, nothing further will be said here about it. But neglect of this topic here should by no means signal that it is of little importance in the overall evaluation of information policy. Indeed, as we shall see in the postevaluative stage, knowledge of agenda setting and legislative enactment of specific information policies can lead us to a better understanding of the success and failure of information policy. One danger that confronts anyone who evaluates policy is that of applying the inappropriate knowledge to the problem at hand. This may happen when economic analysis, for example, is the only knowledge brought to bear on problems of cocaine addiction.

Up to this point, emphasis has been placed on the appropriate knowledge required to formulate *good policy*. Even where personal agendas are involved, it was argued that a rationale for the policy must be put forth that will persuade others to support it. The emphasis placed here on the appropriate knowledge should not be seen, however, as advocating purely analytical policy making, that is, policy making driven solely by the appropriate specialized knowledge. In fact, Lindblom (1980, p. 19) makes the case that:

analytical policy making is inevitably limited—and must allow room for politics—to the degree that:

    1) it is fallible, and people believe it to be so;
    2) it cannot wholly resolve conflicts of value and interests;
    3) it is too slow and costly; and
    4) it cannot tell us conclusively which problems to attack.

The hypertrophied emphasis in this book on the appropriate specialized knowledge is based on the conviction that our knowledge is presently inadequate for the task of any formal analysis and the result is neither good policy nor good analysis. An equally pernicious danger can threaten policy making when politics has left little room for analysis.

What we are left with in the formulation of policy is not a technically designed mechanism that optimizes all the quantifiable variables to the greatest extent possible to produce our desired solution; on the contrary, we are left with the result of political wrangling, power brokering, timing, luck, and to some extent the application of specialized knowledge in the area affected. We must not forget that "the gains from democracy are bought at a cost" (Lindblom, 1980, p. 58). Lindblom further argues that public policies are difficult to understand and to design to achieve predictable results.

Policy Implementation

Once a policy is embodied in law it must be implemented by some executive agency of the government. This setting in motion is a vulnerable stage for the policy. Its success or failure at this point depends on several factors. There are material factors, such as the availability of monetary resources, equipment, well-trained personnel, and facilities. Of more concern in many cases are nonmaterial resources, such as the political climate within the implementing agency, the backing given to the policy by the incumbent bureaucracy, and the clarity and design of the law to be implemented.

Many factors affecting the implementation of policy have been identified. Making a policy does not ensure its implementation in a form intended by the policy makers. At the implementation stage many changes can occur that often are beyond the control of the policy makers and that often cannot be foreseen by even the most astute political veteran. Groups affecting policy implementation include: (1) policy makers (executive, legislative, and judicial), (2) formal implementers, (3) lobbyists or lobby agencies, and (4) the press. The power of each of these groups varies depending on the type of policy.

Policy makers do get involved in policy implementation usually in order to ensure that their policy is implemented in a way consonant with their original intent. They may intervene in the implementation of their policy in different ways depending on their authority or power, and they may take credit (a good policy outcome is a political asset) or disclaim responsibility (a poor political outcome is a political liability) for the actual implementation.

Formal implementers are responsible for implementing a policy made by another set of people, the policy makers, at a different level of government. Formal implementers include administrators within departments, agencies, bureaus, regulatory agencies, and the Supreme Court (Braman, 1988). There are also agents of these implementers, such as governmental bureaucrats at the state or municipal level.

Lobbyists also are active in this stage of the policy process. Lobbying takes place in an attempt to make the implementation of a specific policy work for the lobbyists' constituency. Coalitions of like-minded groups, or separate agencies themselves, may band together to influence the implementation of a given policy. Finally, the press and representatives of the mass media also influence this aspect of the policy process. The well-placed story or the vitriolic editorial can and does have an effect on the implementation of specific policies.

Besides these specific actors, there are other constraints on the implementation of policy: These are organizational structures, such as internal operating regulations, resource allocation guidelines, and the psychological motivations of the implementers themselves. Different organizations have different assumptions about their own operations. Researchers have identified different institutional models of policy implementation: the systems management model that looks at the implementation as a "goal-directed activity"; the "bureaucratic process model" that views implementation as a routine process of continually controlling discretion; the "organizational development model" in which implementation is seen as a participatory process on behalf of the implementers; and the "conflict and bargaining model" that views implementation as a conflict and bargaining process (Nakamura and Smallwood, 1980, p. 54). These organizational models influence the way in which policies are implemented because of the expectations about implementation that are inherent in them. Furthermore, internal rules and regulations often dictate how policies are implemented. The rules and regulations often win adherence over the actual intent of the policy implemented. For example, during the Cuban missile crisis, President Kennedy was often frustrated because the armed forces' standing operating procedures would automatically initiate actions, such as a specific type of alert, that interferred with Kennedy's own executive actions to defuse the crisis.

Any organization, in spite of the policy directive ordering implementation, has to operate within the constraints of its own resources. If a policy requires more resources than the organization possesses, the policy will not be implemented to the extent intended. Finally, psychological attributes of the implementer may influence the implementation of policy.

Depending on the implementing agency and the policy to be implemented, implementation is a value laden process, often with ambiguous instructions on how to deal with contingencies. Lindblom (1980, p. 66) avers that "multiple conflicting criteria are a usual phenomenon in policy implementation."

Information policy faces the same problems in implementation that afflict other public policies. One example of the problems included in information policy implementation is the Executive Department's Office of Management and Budget (OMB) implementation of the *Paperwork Reduction Act of 1980* (P.L.96–511, 44 *USC* 35). The *Paperwork Reduction Act* was intended to (see section 3501):

1. Minimize the Federal paperwork burden for individuals, small businesses, state and local governments, and other persons;
2. Minimize the cost to the Federal government of collecting, maintaining, using, and disseminating information;
3. Maximize the usefulness of information collected by the Federal government;
4. Coordinate, integrate and, to the extent practicable and appropriate, make uniform Federal information policies and practices;
5. Ensure that automatic data processing and telecommunications technologies are acquired and used by the Federal government in a manner which improves service delivery and program management, increases productivity, reduces waste and fraud, and, wherever practicable and appropriate, reduces the information processing burden for the Federal government and for persons who provide information to the Federal government; and
6. Ensure that the collection, maintenance, use and dissemination of information by the Federal government is consistent with applicable laws relating to confidentiality, including section 552a of title 5, *United States Code*, known as the Privacy Act.

To carry out this directive an Office of Information and Regulatory Affairs (OIRA) was established in the Office of Management and Budget. The director of this office was given the general charge to (section 3505(a)):

> develop and implement Federal information policies, principles, standards, and guidelines and shall provide direction and oversee the review and approval of information collection requests, the reduction of the paperwork burden, Federal statistical activities, records management activities, privacy of records, interagency sharing of information, and acquisition and use of automatic data processing telecommunications, and other technology for managing information resources.

A recent Office of Technology Assessment study (U.S. Congress. Office of Technology Assessment 1990, p. 20) points out that in the *Paperwork Reduction Act,* Congress did not "provide guidance on the shape, direction, or even basic philosophy of information dissemination policies that might be promulgated by the OIRA." It was, therefore, left to the OIRA to interpret the law using the guidelines set by Congress, but with leeway in that interpretation. One result of that leeway was the now well-known distinction made between access and dissemination of information and the privatization of information related functions in the Federal government. This implementation of the *Paperwork Reduction Act* has been criticized by Hernon and McClure (1987, pp. 247–250) and defended by the Director of OIRA (Sprehe, 1987). The ensuing debate over accessibility, dissemination, and privatization is a good example of what can result from ambiguous legislative language or the absence from legislation of specific guidance on what turns out to be crucial issues.

In the case of the *Paperwork Reduction Act,* direction was left out because during the formulation of the act other committees in Congress considered passage of another bill, dealing with the printing chapters (Chapters 1–19) of Title 44 *USC* that would have given explicit guidance on information dissemination by the U.S. government. This bill, however, was not passed. (U.S. Congress Office of Technology Assessment, 1990). Congress has yet to provide such guidance.

After the implementing agency decides how to implement the policy, its efficacy will also be contingent on the social setting and the assumption about human behavior upon which the policy rests. Many public policies have gone awry because of improper assumptions made about these factors. One famous case in point is again the *Paperwork Reduction Act*, in which policies apparently beneficial to the internal management of Federal agencies were not consonant with many widely held values concerning information dissemination.

Postimplementation Evaluation

Policy evaluation attempts to answer the question of how well or how close the policy comes to achieving its original goals. The success of the policy will have been influenced by many factors. The postimplementation evaluation must assess not only the original goals, but also the soundness of the rationales put forth for the policy, the history of the formulation stage, the history of implementation and changes to the formulators' original intentions. This is when all the skills of the analyst are used. What went wrong? What went right? Can we assess why things happened as they did?

Policy makers usually engage in the least formal type of evaluation depending primarily on feedback from the policy implemented. Feedback consists of mail, telephone calls, and other types of communications from the policy maker's constituency, which, over time, give the policy maker an impression of how well a given policy has been received by constituents. The policy makers base their actions, to a major degree, on the assumption that satisfied constituents mean reelection; reelection means power, a widely acknowledged goal of politicians. In a less cynical note, Kelman (1987) has argued that public spiritedness also plays a larger role than we imagine in the actions of policy makers and policy implementers.

Policy implementers, on the other hand, have a different approach to evaluation. Like the policy makers, policy implementers have a personal stake in the success of their policy. Their reputation and future depend on such success. Therefore, they attempt to maintain and increase the support of policy makers for the policy implemented. Because of this, the type of evaluation coming from implementers is apt to be biased in favor of the policy. Policy implementers can influence the evaluation of policy in several ways: They can (1) filter information about the success of the policy, (2) mobilize support from groups affected by the policy and urge positive feedback to the policy maker, and (3) use resources of the program implemented to gain support from affected groups.

The political stakes are high for both makers and implementers in the evaluation process. Both groups of actors will naturally attempt to put the best face on any policy with which they are involved. Because of this bias, evaluations of policy emanating from these groups may not be the most accurate indication about the success of a specific policy. Another group of evaluators, technical evaluators, can sometimes provide the corrective that is needed for biased evaluations.

Technical evaluators are usually hired by policy makers or policy implementers in order to provide an objective evaluation of a specific policy. Of course, because of the financial arrangement involved, these evaluators are subject to pressure from their benefactors to examine a policy outcome in the most favorable light. These technical evaluators also must maintain their reputation as professional evaluators. They do this by providing an evaluation that is as objective and unbiased as possible. In order for a policy to be evaluated as objectively as possible, it must meet certain basic tests (Nakamura and Smallwood, 1980, p. 72):

- Policy goals are stated clearly;
- These goals are precise enough to be measurable;
- Implementation activities are directed toward achieving these goals;
- Objective measures that relate implementation activities to goals exist or can be created; and
- The data necessary to verify these measures are available.

The main problem usually encountered is that the first test is not met. Often the goals of the policy are stated in language that cannot be measured. For example, in one paragraph of the statute that set up the National Science Foundation (P.L. 81–507, sec 3[a][5]), one goal was to "foster the interchange of scientific information among scientists in the United States and foreign countries." On reflection, one will realize how difficult it would be to assess whether this goal had been met. In the case of information policy it may be even more difficult than we think. If our theories about information flows are weak or nonexistent, we cannot focus on the evaluation of the analytical design of the policy but only on the institutional design of the policy. Political critique will overwhelm analytical critique and become the dominant mode of discourse about the type of policy in question. This has occurred in many areas of information policy.

Analytical design refers to those processes specified by a policy on account of some nonpolitical analysis of the problem. Institutional design means those organizational specifications of a policy that are mandated by the particular political system in which the analytical design is being carried out. In economics we know that taxation will limit the amount of money available for consumer spending and may have an effect on the health of the economy as a whole. So as a matter of analytical policy we decide to alter the amount of taxation. This is analytical design. The questions of which unit of government can order that this be done, how the taxes will be collected and the penalties for nonpayment of taxes relate, for the most part, to institutional design.

This is not to say that information policy is the only type of public policy where the specialized theoretical/analytical knowledge of an area is not robust. Many others surely are. Welfare policy is another example.

POLICY TYPES

One way policy analysts have attempted to broaden the analysis of public policies in general has been to examine them and classify them into types on the basis of their fundamental goals. Lowi (1972, p. 7) has described four types of policy that are identified by their functional goals. These types are distributive, redistributive, regulatory, and constituency-based. His purpose (Ibid.) in describing public policy in this way is to enable political scientists to "develop criteria for policy choice in terms of predicted and desired impacts on the political system, just as economists, biologists, and the like attempt to predict and guide policies according to their societal impacts."

Distributive policies are those that are "characterized by the ease with which they can be disaggregated and dispensed unit by small unit, each unit more or less in isolation from other units and from any general rule. 'Patronage' in the fullest meaning of the word can be taken as a synonym for 'distributive'" (Lowi, 1964, p. 690). Regulatory policies attempt to control the action of a group of persons or a corporate body by allowing or prohibiting behavior. The decision "involves a direct choice as to who will be indulged and who deprived" (Ibid.). Examples are allocation of television channels and truth in advertising.

Redistributive policies confer benefits, much like distributive policies, but they are different because they simultaneously take away benefits from other groups. The fourth type of policy—constituency-based policies, sometimes referred to by others as self-regulatory policies—are the most difficult to describe and characterize. According to Salisbury (1968, p. 158) they "also impose constraints upon a group, but are perceived only to increase, not decrease, the beneficial options to the group. " Lowi (1972) includes reapportionment or setting up a new agency as examples. These are distinguished from redistributive policies by the type of group immediately affected. With constituency-based policies, the political party is the beneficiary. With redistributive policy, nonparty groups are immediately affected such as reserve controls of credit, progressive income tax, and social security.

Characterizing policies in this way helps us to understand what the policy actors perceive to be the functional goal of a policy. If we know the functional goal of a specific policy as perceived by the policy actors, we may be able to predict how the policy will be influenced by different policy actors, both public and private. Also, given the historical circumstances at the time of the policy making, the probable success of adopting a policy can be assessed.

For our purposes it is enough to realize that political scientists have attempted to classify policies according to functional goals in order to predict their effects. When applying these four classes to information policy, as Linowes and Bennett (1986) have shown, it is clear that information policy-making and implementation may be more difficult to understand than are traditional public policies that can be easily classified by Lowi's four-part typology. This difficulty arises because, while many different types of policies are subsumed under the rubric of information policy, it is not clear that their common characteristic of information coincides with each of the policies' functional goals. Further, certain information policies, such as privacy policy, may belong to another as yet unnamed policy type as suggested by Linowes and Bennett (1986).

In speaking about these four policy types, based on functional goals, the functional goal was that perceived by the policy actors. Assumptions about the perceptions of policy actors, as well as their motivations, determine, to a great degree, the analysis of a given policy. Analysts' assumptions about the perceptions of policy actors and the actors' behavior can be described as models of policy making. Earlier in this chapter brief mention was given of the so-called "rational-actor model." The next section elaborates on this and other models developed by political theorists that attempt to predict the actions of policy makers.

MODELS OF POLICY MAKING

Several models attempt to explain the way people in organizations make decisions. Political scientists continuously search for the best model to apply to a given public policy. Here only three such models (the rational-actor model, the bureaucratic model, and the garbage can model) will be briefly examined. The description of these three representations demonstrates the wide range of such models.

The rational-actor model assumes that policy formulators and implementers are rational and act unencumbered by external events. The main guide for action and decision making is rationality. Hence, the decision making process involves the following steps: (1) identification of the problem, (2) consideration and description of all facets of the problem, (3) offering several solutions for the problem based on the constraints delineated in the second step, (4) outlining the advantages and disadvantages of each solution, and (5) choosing one of the solutions based on the assumption of achieving maximum benefit. The model carries with it some assumptions: There is enough time to carry out the process outlined; no interfering variables arise during the decision making process; and once the decision is made for action, the desired result will be achieved.

Although this model of organizational decision making is often used as a straw man by analysts, the assumptions that make up the model can still be seen in many public media accounts of policy choice and tragic events. When this model is applied, one of two things happens. Either failure of the policy is attributed to the failure of the rational actors to consider all the alternatives and make the right choice (the events leading up to the bombing of the U.S. marine barracks in Beirut in 1982 is but one example), or the analyst is bewildered about why the policy implementation did not turn out as planned. The contention in this latter case is that the policy implementers followed the dictates of rationality but still were foiled.

The second decision making model, the so-called bureaucratic model, assumes that in spite of the intent of those formulating the policy, once it moves to the implementation stage, certain phenomena affect the policy in possibly undesirable ways. The person using the bureaucratic model assumes that the bureaucracy is essentially a rule-governed system. People within the system make decisions on the basis of clearly established rules and modes of conduct. Any policy that is to be implemented by a bureaucracy must be transformed in such a way that whatever the intent of the policy makers about the implementation of the policy, the bureaucratic rules of the implementing organization will be more powerful. In this type of implementation, rules, not policy goals or rationality, carry the day.

The third decision making model, the garbage can model, is at the other end of the spectrum from the rational actor model. This model applies to decisions where the goals may be far from clear and the methods and constraints on implementation are assumed to affect the final policy outcome. In this model the decision making process is pictured as a garbage can into which policy goals, organizational rules and constraints, the "right climate," and other often unexpected variables are thrown together. The resulting policy outcome is often unpredictable and because of the ambiguous nature of the policy goals themselves, often unrecognizable.

The reader of a policy evaluation should be aware that analysts have certain assumptions about how decisions are made in government. These assumptions are often not explicit, but they certainly affect the conclusions of the policy evaluation. One way of testing the credibility or legitimacy of a given evaluation is to ask what assumptions about organizational choice the writer makes. If the assumptions are in doubt then the entire evaluation may come into question. Finally, one related area of any given policy may be more difficult to ascertain than the writer's assumptions about decision making— the assumed values of the evaluator.

VALUES INHERENT IN POLICY EVALUATION

Marney (1961, p. 17) has stated that "all institutional loyalties are value judgments. Institutionalism becomes a structure native to prejudice precisely because the institution exists to mark the edges of a valuable." All analysts have some type of institutional affiliation. These affiliations and loyalties inhere in the analyst's judgment about a specific policy. For example, the analyst who works for the Environmental Protection Agency (EPA) can be expected to have a different analysis of an EPA policy than would the analyst who is employed by a waste disposal firm. These types of institutional loyalties are strong and probably the most visible. But other types of institutional loyalties are stronger and are often more difficult to perceive —that is, the institutions of law, economics, politics, science, technology, and religion. These types of institutions, as much as those mentioned earlier, have more power and influence on the conduct of policy analysis and on the conclusions reached as a result of that analysis.

For example, the scientist arguing for the absence of any control on scientific communication will certainly and obviously be in conflict with a politician who claims that such unimpeded free access will damage the security of the United States. The creationist's claim that his or her beliefs are scientific will have an argument from a scientist who disputes that claim, or from a district judge who disputes it on legal grounds. These are cases where the conflict between the values of these institutions are easily discernible. But in other cases the conflict may not be readily perceivable because the reader adheres to the values of both institutions. For example, there has been much media attention over the past decade about the technological race between the Americans and the Japanese. Vogel (1979), in *Japan as Number One*, makes four recommendations for catching up with and surpassing the Japanese in high technology. One of these urges a deemphasis on individualism in this country for the good of the larger whole. It may be necessary, suggests Vogel, to curb civil liberties in order to achieve our goal of being number one.

Although this may be oversimplifying Vogel's argument, we are faced here with a conflict in value institutions in which most of us are caught. On the one hand is the institution of economic security and supremacy, and on the other is the American ideal of liberty guaranteed in the Bill of Rights. This type of value-laden analysis is more difficult to deal with because we have allegiance to both institutions of value. It is in this area of public policy analysis, the resolution of ambiguity in decision making and policy choice, that we face our greatest challenges.

SUMMARY

This chapter has explored the political context in which information policy is made. Policy making, implementation, and evaluation are the three widely recognized stages of the policy process. Policy theorists constantly try to refine their understanding of this process in order to predict the outcome of specific policies. This desire to predict has led them to classify policies into different types, assuming that different factors will affect the outcomes of different kinds of policies. Similarly with decision making models, the theorist's aim is to get some kind of predictive handle on specific policy tasks.

The more accurate our ability to predict social and political behavior under given conditions, the better is our social science knowledge. This knowledge is often multifaceted and includes not only knowledge about the policy process in general but knowledge underlying the rationales used to achieve specific functional goals (e.g., increase taxes and reduce disposable income). If our knowledge of a specific policy area cannot be substantiated or be used to evaluate rationales for achieving specific functional goals, then we may call this knowledge weak.

The weakness or strength of our knowledge concerning information policy will be evident when we isolate rationales for specific goals and evaluate their soundness. Does the rationale given match what we know about the behavior of people and information in the context given? Will the expected goal be met if the recommended action is taken? If our knowledge is inadequate for this task, then the likelihood of achieving the goal of the policy is questionable.

It can clearly be seen that the soundness of information policy and the state of our knowledge about the behavior of people and information are interdependent. Without the appropriate knowledge, the policy will not be sound; without the making of information policy and the appropriate evaluation of our successes and failures, our knowledge will languish. Appropriate evaluation would include, but would not necessarily be limited to, determining the soundness of the rationales given for specific actions taken under given conditions and evaluating the knowledge used at the time to validate or undermine those rationales. In this way we will be forced to examine not only the political factors included, but the effects of the policy and the degree to which those effects were predictable or not.

Part of the evaluator's task is to determine to what extent our knowledge about information phenomena is adequate. Much of the work that has already gone on with information policy analysis has focused on its formation and implementation, but little emphasis has been placed on the theoretical knowledge necessary to make the evaluation of information policy as robust and complete as possible. This is not just a problem confined to the United States, but an international problem as well. Its solution, therefore, is not limited to U.S. solutions, but must include information policy problems and their solutions worldwide.

CHAPTER 2

INFORMATION POLICY IN AN INTERNATIONAL CONTEXT

Chapter 1 pointed out that the formation and implementation of information policy, or any public policy, are complex and often unpredictable undertakings. Complexity and unpredictability increase as we involve greater geographical areas, more political systems and different channels over which an information policy is intended to exert control. It has become axiomatic that information policy made in one nation will have effects elsewhere.

This insight, however unremarkable it may seem, can lead us either to an extremely optimistic view or an extremely pessimistic view of information policy making and its potential success. The optimist may claim that society is essentially a cybernetic system, governed by controls and feedback loops. Our task is to gather as much information about the area over which we want control, construct models describing how the many variables interact and design a policy to meet our goals. The optimistic view is perhaps best represented by Wiener (1950).

The pessimist, on the other hand, is skeptical about our ability to gather enough data to predict with any reasonable assurance of success what the results of our actions would be. The pessimist doubts that society works like a servomechanism even though many believe it to do so. Our technology, rather than being a heaven-sent tool, is actually our foil. All our models can do is to approximate in a grossly inaccurate manner the behavior of human beings and their probable responses to certain conditions. We may believe that we can fix our problems, but our very act of fixing creates new problems and situations that were not only unforeseen, but were unpredictable. The pessimistic view is represented by Ellul (1964).

I fall somewhere between these two extremes. Neither persuades me fully, however, either of one's ability to control society or of the absolute inability to effect some positive change on the world around us. What propels me into this endeavor is what many regard as the naive notion that it is incumbent upon us as responsible human beings to attempt to control phenomena in an effort to help other human beings. For example, economic development, I believe, can help alleviate some human suffering. Information policies, inter alia, can aid economic development. Therefore, I believe it is our responsibility to others that should encourage us to formulate and implement the best information policies possible. The burden of this book is that knowledge about information policies and their effects is necessary to formulate such policies and we must discover where our level of knowledge now stops and where our ignorance begins. Abdication of these decisions to experts that claim to know what is good for us without admitting their ignorance of the principles underlying information policies is intellectual sloth and cowardice.

On the other hand, I am also hesitant to pursue this approach knowing that such knowledge, if sufficient to formulate and implement effective information policies, could also be used to harm human beings, by thwarting economic development or by restricting humans' ability to acquire and use information, thereby limiting their autonomy and eroding their dignity.

Having admitted this moral dilemma, I would like to examine some selected facets of information policy on a global scale. My purpose here is not to provide a comprehensive view of such policy making, but primarily to enlarge our sphere both of the probable effects of our policies and the variables that affect our policy making efforts. Following this expansion of our perspectives, I will narrow the focus of concern first to scientific and technical information (STINFO) policies (Chapter 3), and then to a close look at one specific STINFO policy, *Scientific and Technical Communication* (1969) (Chapters 6-8).

At the present time, the commonly held view is that information policy in general has been driven by the need for catalyzing economic development and increasing economic productivity. The level of development and productivity directly affects the degree to which a nation can solve internally its own economic and social problems and prevent cultural and societal stagnation. Although we may disagree with this common characterization of the catalyst for information policy, it is undeniably a widespread notion among policy makers. This is not to imply that we should ignore other reasons for pursuing better information policies, or to seek more human-oriented reasons for developing them. It is simply a statement about the current state of affairs.

For the purposes of this book, however, the problem of what knowledge determines a nation's information policy is paramount. Whatever we believe the *raison d'etre* for information policy is or should be, or to what degree we are cognizant of our own limitations in controlling national or global information flows, we must possess some knowledge about the purported effects of carrying out a specific policy. Before we can apply any knowledge to solving information policy problems, however (problems engendered in part by the desire to acquire and use information for development and for international competition), we must first examine the international context in which current information policy is formed. I say "must examine" because of the acknowledged interdependence of the world's economies and social and cultural systems.

In the previous chapter the focus was on the United States' political system and its complexity. We were concerned there with the context of the institutional design of the political system within which policy is made. A similar complexity of political institutions exists in each and every country attempting to formulate a national information policy for itself. In examining the political or organizational context in which policy is made, this book has purposely diverted attention from a solely technological perspective on the information policy problem. This chapter includes technology among several other aspects of the process through which information policy is formulated and implemented. The problem with the institutional design of information policy is compounded in countries that do not already possess the technological superiority of the United States. To be sure, the United States also has to contend with many of the problems outlined below, but it can do so from a position of strength and superiority. There is little anxiety about catching up with the rest of the world. Instead, any national angst in the United States regarding information technology is the product of the need for remaining ahead in technological superiority. For most other nations, however, the motivation is not to stay on top and stay in control, but to catch up and resist being left behind. In these countries, policy is formulated and implemented, not primarily on the basis of a knowledge of all the systemic effects of such policies, but primarily on the belief that if it worked for another developed country or a country that has already achieved technological and economical superiority (e.g., the United States and Japan), it will work for them. This belief is based on experiential knowledge, a knowledge not necessarily reliable, verifiable, or presently codified. As Von Laue (1987) has shown, the wholesale import of a system, be it political or technological, into another country that does not have the same cultural milieu or level of social discipline can, but usually does not, succeed. This condition notwithstanding, however, there continues to be an unabated thirst for parity with developed countries in the realm of information control and use. This desire is only occasionally questioned from social, cultural, and religious viewpoints.

The discussion that follows, therefore, is intended neither to survey all world information systems nor to catalog the issues before us. In this regard see Surprenant (1987), Eres (1989), Moohan, Morton, Rimmer, Romano, and Burton (1988), and Mowlana (1985). Instead, the purpose is to survey the complexity of the environment in which each nation has to make national information policy. To do this, the chapter examines several reasons for establishing information policy, some methods employed to control information flows, and various issues of justice and rights that affect and are affected by information policy making. This environment must include a global awareness of how a nation's own information policy affects and is affected by other nations and by other unpredictable factors beyond its control in the dynamism of global existence. One broad statement is admissible, however, concerning some of the forces that have made it necessary to think about national information policy and its global implications. We are concerned with information policy because we wish to control our environment and forestall or minimize disasters, whether those disasters be natural, environmental, economic, military, or cultural. This desire to control the world beyond our own local communities has developed alongside our ability to extend our influence beyond our local communities in the economic and social spheres. This need to control an ever-widening area beyond our local communities has been called the *control crisis*.

THE CONTROL CRISIS AND THE RATIONALE FOR ESTABLISHING INFORMATION POLICIES

Beniger (1986) makes the astute observation that the information society was precipitated by the control crisis. He describes the crisis in control that came about when local markets were no longer isolated from each other. This isolation, formerly enforced by the barriers of the lack of transportation and long distance communication, was eliminated by industrialization. With these barriers eliminated, the distribution of goods and services could then extend beyond local markets to national and global markets. Direct communication between producer and consumer, which had previously served to regulate production in the local market, is no longer possible. Durkheim (1893, pp. 369–370) describes this breakdown of traditional, local markets, which he calls segmental markets, to form national and global markets:

> Insofar as the segmental type is strongly marked, there are nearly as many economic markets as there are different segments. Consequently, each of them is very limited. Producers, being near consumers, can easily reckon the extent of the needs to be satisfied. Equilibrium is established without any trouble and production regulates itself. On the contrary, as the organized type develops, the fusion of different segments draws the markets together into one which embraces almost all society .... The result is that each industry produces for consumers spread over the whole surface of the country or even of the entire world. Contact is then no longer sufficient. The producer can no longer

embrace the market in a glance, nor even in thought. He can no longer see limits, since it is, so to speak, limitless. Accordingly, production becomes unbridled and unregulated. It can only trust to chance . . . . From this come the crises which periodically disturb economic functions.

The crisis in control precipitated the control revolution, which fundamentally was an attempt to maintain control over the extended markets. The control revolution was fueled, in turn, by the increasing sophistication of transportation and communication technologies. Control was subsequently sought not only by producers, but by anyone or anybody that wished to remain in control. Political, business, social, educational, and religious leaders all sought to carry out their societally appointed tasks, to maintain their power, or to gain access to power by controlling information of various kinds. Information policy is the tool by which this control is maintained or lost, by which power is shared or retained. Various promoters of a national information policy may, therefore, have different political motives for a given policy. Their suggestions regarding what the policy should be is a genuine determinant of the distribution of power in a society.

It is no wonder that on the international scene, individual nations have to be concerned about the internal information policy of their country—who controls the broadcast media and the press, the amount of independence given to them, who is literate or illiterate, what scientific literature is classified or unclassified and a host of other related issues including what controls, if any, should be placed on information flows into and out of their country. Controls of one sort or another will be instituted to protect or promote indigenous commerce or culture. The unique aspect of information policy is that, unlike some economic or social policies, any national information policy affects or is affected by what occurs elsewhere in the world. This has come about because of the control revolution and the proliferation of information technology to master the crisis in control. In an inescapable way, all nations are part of a global system of information flows that affects their stability and way of life. Information policy is a tool as important as economic, energy, educational and foreign policy because it is entwined with all attempts to exert power in a society.

PURPOSE OF INFORMATION FLOWS
AND THEIR SCOPE

An information infrastructure (institutions employing information technology to produce, store, analyze, and disseminate information of all kinds) is most often seen as central to the economic well-being of a nation and primary in its international standing. Countries do not often realize that they have not progressed as far as they should have economically until they look about the world and find themselves lacking. For example, Mandeville (1987, pp. 36–37) called for the recognition of the importance of information for the Australian economy and suggested five items to be placed on the national economic and social agenda:

> First, appropriate action to shift policy from almost total emphasis on the production of what is seen as "real" wealth, that is wealth in the form of tangible goods emanating from the tangible-goods sectors. Although most wealth no longer comes from these sectors, or in the form of tangible goods, economic policy still seems to assume that it does. Most wealth is now created in intangible form and most of that intangible wealth is information. Economic policies which fail to recognize this, which continue to assume that activities producing tangible wealth are the core of the economy on which all else is dependent, are likely to be irrelevant to the means by which most wealth may be produced and most employment created.
>
> Second, the market system copes best when problems are well defined, information standardized, and when there is a low level of uncertainty. Where these conditions are not met, there is need for the invention of new forms of organizations—new ways of organizing social and economic activity.
>
> Third, people—not capital, not energy, not natural resources not materials, but people—and the communication of information among them are the crucial inputs in today's economic activities. High-technology firms illustrate this point well. Perhaps our abundant natural resources distract our attention from this point. Japan's only resource is the Japanese, yet this Asian nation is probably the most successful one in the information era. Personal and organizational efficiency require a capacity to learn, ask questions and use information—in other words, the capacity to initiate change and adapt to it . . . .
>
> Fourth, in Australia institutions, businesses, policy makers, and the public need to develop more skill in scrapping what is outdated. In the economic sphere, exposure to international competition is a great help in this process. More basically, our concepts of work, the economy and society need updating.
>
> Finally, perhaps a national information policy is needed to focus attention on these new realities.

Mandeville (1987) focuses on commercial information and its role in the countries' economic well-being. He emphasizes the importance of information within the country so that the country can become more economically productive and expand markets overseas. He sees himself as a modern-day prophet whose mission it is to awaken his countrymen to the dismal comparative situation in which they are mired.

Gray (1988), writing on information policy in Malaysia, emphasizes a more fundamental problem, one often faced by lesser developed countries. While recognizing the importance of information for social and economic development, the information infrastructure of the country must first be built up. For Malaysia, both governmental and public recreational information resources must be created, collected in libraries, and documented so that they can be exploited. In short, as Wijasuriya (1988) claims, the crucial need is for providing for literacy and basic information provision. These types of concerns about the basic information infrastructure, however, can be further complicated by religion or ideology.

For example, Sardar (1988, pp. 118–119), in writing about the national information systems in the Muslim world, laments:

> The establishment and successful development of a system of communication of science within the *ummah* presupposes the existence of a viable information structure. While a few Muslim countries have certain types of national information services, national libraries, and information centres catering for the scientific community, fully developed and integrated information infrastructures are conspicuously absent. An integrated information infrastructure—linking national libraries, computer-based data centres and archives and public and rural libraries, community information referral centres, reference and scientific information centres—is like the nervous system of a country. Knowledge flowing through this system is the vital life fluid which stimulates renewal and growth. Without it, Muslim countries are like a body without a nervous system: inert and static, unable to feel and adjust to change.
>
> Without a developed and integrated information structure the smooth and efficient functioning of a modern government is impossible . . . . It is the availability of a wide range of information on a variety of important subjects to a nation which makes it possible to comprehend the true magnitude of problems, set future goals, and choose wisely from a range of options. In choosing between alternative courses of action and policy, people require information on the possible outcome of this choice, as well as on techniques to achieve them and the assumptions behind them. Thus information plays an important role in every aspect and stage of planning administrative and policy work.

But, as discussed in Chapter 1, the possibility of making rational policy decisions based on an adequate supply of information is a questionable enterprise. We may like to think that the rational-actor model of policy making is possible, even desirable, but its shortcomings are all too well known. We would like to think that we can gather all relevant information bearing on a problem so that we can assess the risks, costs, and benefits of various alternatives. Decision making rarely follows this ideal path. This caution about the ability to make rational policy notwithstanding, Sardar (1988, p. 164) argues for a specific filtering of information because

> At present, there seems to be at once too much information and too little information: there is abundance, indeed superabundance, of worthless, obnoxious, even manipulative information, and a scarcity of relevant, high-quality information . . . . There is a need, on the one hand, to fight the degradation of

our moral and social environment, and, on the other, to develop local and international avenues and channels of indigenous self-expression.

He is speaking from a powerful religious tradition that puts Islam first and economic and development concerns second. This choice of priority of goals for information policy affects the way a nation uses internationally available information. To what extent this information affects the social fabric of a country (Borgmann, 1984), and in what ways this information may be controlled according to the indigenous desires of a specific country's ruling elite are two questions not yet having clear answers. The question of an abundance of information and the need for its filtering, whatever the purpose of the filter, is also raised by Cronin (1987, p. 92).

In general, at the present time, the primary push to develop an information policy, especially a policy that will take advantage of international information flows, is seen as an economic one. Cronin (1987, p. 91) maintains that one common thread running throughout most nations' information policy is "an awareness that successful economic functioning is inextricably linked to the effective management of information resources at both the social and personal level." He claims, at least where developed economies like Japan are concerned, that "support for the information industry sector is crucial to the development of a robust national industrial policy" (p. 95). Once the crucial importance is recognized, whatever the priorities of the country involved, then the questions arise concerning what technical tools will be needed and what their effects are; what legal, economic and political tools will be needed and what their effects are; what multinational bodies must be appeased or turned to for assistance; what political and cultural barriers will inhibit or promote such policy making; and what ethical issues are entwined in the process of making these decisions. The following discussion is not intended to be exhaustive or comprehensive, but is provided in order to sketch out the international context in which information policy is made. The next section emphasizes the complexity and interrelatedness of some of the major aspects of information policy making that are now often discussed in an isolated way in the literature.

TECHNICAL CONSIDERATIONS

Information can travel across national boundaries by several physical means. The information can pass by word of mouth, by post, by broadcasting, or by telecommunications networks using voice data between persons or digitized data communicated between computers. It can also be transferred as an object, in the form of a technological device that recipients can examine in order to decipher its inner workings. This process is often referred to as reverse engineering.

These levels of complexity of the channel of information from personal communication to computer communication increasingly demand greater standardization in order to process the information effectively and intelligibly. Beniger (1986, p. 15) describes this standardization as preprocessing; that is, "control can be increased not only by increasing the capability to process information but also by decreasing the amount of information to be processed." This preprocessing and standardization can take many forms, but it must be employed in an ever greater sophisticated way as we move from personal communication to computer digitized communication.

Personal Channels

At the first level of complexity a common language is needed. If people communicating with each other do not speak and write the same language, they will not be mutually intelligible. Issues related to the exportability and importability of information, no matter what the channel, revolve around this concern. For example, in some developing countries, access to information not communicated in indigenously spoken and written languages is limited to those elites who can speak other languages. As far as export is concerned, unless a writer can communicate in a language commonly used around the globe, the information will not be accessible and probably not considered valuable. Lim (1989, p. 21) reports that databases created in developed countries often do not cover material published in a developing country nor do they include topics relevant or of interest to those LDCs (Lesser-Developed Countries). Furthermore, Lim (p. 27) claims that another barrier to the use of information from developed countries is the use of only a few scripts to convey the information. Many LDCs use nonroman scripts for their languages. The information conveyed by these scripts is unavoidably, but arrogantly, considered less valuable or reliable by those speaking languages more universally used in the technological enterprise, such as the English language.

This idea of the credibility of information published in uncommonly spoken languages or in journals that are not widely available is not limited to lesser developed countries, but is applicable to all languages and cultures. Here the credibility of the information is not challenged on ideological or religious grounds, but simply on grounds of accessibility or perceived worth. This language problem, however, is not only limited to linguistic groups per se, but also to specialty languages of disciplines and fields of study. Mikhailov, Chernyi, and Giliarevskii (1984, pp. 25–29), for example, describe intralanguage and semiological barriers in scientific communication when, for example, an algebrist may not understand an article by a number theorist, even though the article is written in a natural language spoken by both mathematicians. Another example of such barriers is the case of an educated scientist in one discipline not understanding a general theory in another scientific discipline. Policy makers must be aware of these barriers in order to overcome them. In order to do this information policy might overlap with educational policy and the establishment of foreign language training programs, or with science and technology policy in the establishment of translation centers.

Both of these solutions require expenditure of scarce resources that may be beyond the capacity of the country to provide. Instead, resources may be diverted to the establishment of a flashier symbol of technological superiority, such as a computer network. Once established, however, there is a lack of trained personnel to process, analyze, and interpret the information conveyed over the seemingly more attractive communication channel.

Postal Channels

At the next level of complexity, postal service, standards, such as the size and weight of envelopes, will determine whether or not something may be sent by post or how much it costs to do so. International postal agreements had to be set up regarding such standards, as well as the honoring of international fees for mailing and the reimbursement of receiving and sending countries for services rendered.

Broadcasting

At yet another level, broadcasting, one important technical tool to be controlled is the frequency spectrum. Only a finite number of broadcasters can broadcast at any one time in a given area. This is because of the physical limitations of the broadcasting bandwidth to send an intelligible signal. In addition, each part of the spectrum reserved for sending a signal must be separated by another, unused part of the spectrum so that two signals sent on adjacent channels will not interfere with one another. This is a relatively manageable problem within one country with conventional technologies, such as radio. But as soon as satellite communication became a possibility, there arose the problems of both a finite number of satellite positions in the sky and a finite number of channels for each satellite. Many Third World countries that had not developed satellite communications demanded that they be given access to some of the spectrum, even though they had not yet developed the technological capability to use it. A similar conflict has arisen with the demand for the limited spaces in geostationary orbits for communication satellites. For example, the Kingdom of Tonga in the South Pacific lay claim to sixteen geostationary satellite slots. It did this through a loophole in international law that allowed allocation of these slots to either individual sovereign nations or to Intelsat (a 119 nation consortium that already provides most international satellite service). Tonga's economy is largely based on fishing, coconuts, and foreign aid. Therefore, many doubt the ability of this nation to use the slots, suspecting that Tonga's goal is to make a profit from them by leasing the slots to the highest bidders (Andrews,1990). The WARC conferences have been devoted to resolving these kinds of problems (Surprenant,1987).

Computers and Telecommunications

Computer and telecommunication standards present another level of complexity. Here the standards become so numerous and technical that the average policy maker does not have the expertise to understand them. But some of these standards are fundamental to communication. For example, the ASCII character set is necessary to communicate text through computers linked in a network. The MARC (Machine-Readable Cataloging) standard allows the transference of bibliographic data among libraries and bibliographic utilities. Specialized standards for telecommunication interfaces, designed so that the signals transmitted along a specific set of wires will be connected to the same set of wires in a receiving computer, occupy a crucial place in the communication chain. In order for bibliographic networks to transfer data to one another in real time a standard known as the Open Systems Interconnection was formed. This is the foundation for the SNI (Systems Network Interconnection) protocols (Denenberg, 1988). This small catalog of selected standards is given to underscore the point that a great many techniques must be perfected, a great deal of coordination must take place, and a great deal of knowledge must be gained about the technical functioning of communication systems even before they can be controlled in the ways most beneficial to humans. When policies are formulated and implemented, many assumptions are made about the working of technical systems; the complexity of these assumptions is often invisible to the policy maker. Not every policy maker must become a technical expert in order to formulate good policy; nor should only engineers and scientists, because they possess the requisite technical knowledge, become policy makers. Rather, this is simply a statement about the complexity of the communication channels and how we deal with the complexity.

   In countries that are just introducing information technologies, the technical complexity alone may be overwhelming and may limit the pace of that introduction. For the most part those of us in the West take this technology for granted because its complexity is invisible to us (Borgmann, 1984), or we assume that its functioning does not have any effect on the policy area in question. This caveat notwithstanding, information policy is made, at times with an ignorance of the technical complexities of these policies, or with an ignorance of the technical feasibility of a proposed policy. In addition to these technical tools of control, several political, legal, and economic means are used on the international scene to obtain information from other countries and to restrict information from crossing a country's borders.

POLITICAL, LEGAL, AND ECONOMIC CONSIDERATIONS

Political, legal, and economic considerations are usually brought together and discussed under the rubric of transborder data flow (TDF). Limiting the discussion to the traditional categories of TDF, however, would exclude some powerful tools that can be used to promote or impede communication. These tools cannot be applied along the same hierarchy as were technical tools. Generally, they may apply to individuals or to groups of individuals or the products of individuals and occasionally overlap or include technical tools.

An information policy may control information flows among individuals in several ways. For example, the training of students in scientific and technical disciplines and foreign languages, as well as providing for study abroad opportunities, is a basic, but long-term, tool of information policy. A bilingual person can read and understand information in another tongue and communicate that information back home. Many nations of the world stress bilingualism and urge their students to study abroad, often in Western Europe or the United States. Many policy makers in the United States have become alarmed at this trend, citing the high proportion of U.S. doctorates in mathematics that have been awarded to foreign nationals over the past several years (Browne, 1990).

A nation unwilling to share its scientific and technical literature with other countries might attempt to limit the number of foreign students studying there, might restrict attendance at scientific and technical meetings to citizens of their country, or may try to limit the distribution of scientific and technical literature over which it has control (Relyea, 1985; *Scientific Communication and National Security*, 1982).

Another way to control information flows is by controlling artifacts that themselves embody information, such as computers, remote sensing devices, cryptological software, and various electronic devices. These controls are usually in the form of some export prohibition, such as the United States' *Export Control Act*, or tariff and other trade barriers that impede the flow of information embodying artifacts. PTT (Post, Telephone, and Telegraph) regulations also control international flow with a monetary regulatory mechanism, such as price of international mail or transborder data flow charges for international telecommunication messages.

As was the case with the technical considerations, behind these political, economic, and social tools lie many assumptions about the effects of the tools' application. For example, tariffs are meant to protect national industries from foreign competition. However, the use of tariffs can also be damaging by lulling the native industries into a competitive complacence whereby their goods lack the quality of similar goods available abroad. But what of the application of some of the noneconomic tools? The governmentally placed restriction on the freedom of access to scientific and technical literature may or may not prevent foreign nationals from acquiring that information. Such restrictions, however, can also restrict that information to scientists and engineers of the restricting country, thereby artificially shrinking the knowledge base of the country itself and damaging it more than the perceived threat of foreign acquisition of that knowledge could have brought. Our policy making in this area is sketchy at best and always of a divided mind. The basic issue is the assumed effect of information on the acquirer of that information. All across information policy issues, from pornography and privacy to global broadcasting systems and the dissemination of scientific and technical literature, one fundamental concern is the assumed effect of a given body of information on the receiver of that information and the value judgment of whether that effect is "good" or "bad." U.S. information policies are based on these assumptions and the United States has joined with its friends in other nations to preserve information flows that they perceive to be beneficial to them, and to impede information flows that they regard as harmful to them. The results of these mutually perceived value concerns have been the establishment of international bodies that promote certain types of information flow both within countries, to develop national information infrastructures, and among countries, to aid mutual information exchange.

MUTUAL RESPONSES

In employing any of the tools discussed thus far, this chapter has omitted any consideration of the assumed response of nations affected by another nation's policy. The imposition of a restrictive tariff may elicit a condemnatory diatribe against the imposing country in an international consultative body, such as the World Trade Council. In other cases, the response could be a retaliatory tariff. Depending on the type of government the affected nation has, the response to another's policy may be swift, in kind, or deliberative and conciliatory. National domestic information policy has thus become an element of foreign policy.

Influence of Multinational Bodies and Programs in
Global Information Flows

Another consideration is the influence of multinational bodies in global information flows. The United States may be a part of an organization that has developed programs relevant to information. Our national policy may contribute to or impede that program. Several examples of these considerations follow.

Unesco. Unesco (The United Nations Educational, Scientific and Cultural Organization) has taken a leadership role both in giving developing countries more voice in mass communications and in facilitating access to global information resources through the development of national information systems. It has done the former through promulgation of the New World Information and Communication Order, and the latter through several programs, three of which will be discussed here. These programs are UBC (Universal Bibliographic Control), UAP (Universal Availability of Publications), and UNISIST. These programs have had the dual purpose of developing information agencies in countries that lack an information infrastructure and of providing access to publications on a global scale.

New World Information and Communication Order (NWICO). The inability of developing countries to cope with the technological hegemony and economic and communications power of the developed world was the primary reason for the call to a New World Information and Communication Order. It was encouraged in part by the MacBride Report (MacBride,1980), a multiauthored document that detailed the many grievances of developing countries with regard to worldwide communications and their effect on these countries. Surprenant (1987, p. 49), quoting Masmoudi (1979), presented an outline of the fundamental complaints:

- A flagrant quantitative imbalance, in information, between North and South (hemispheres)—South overwhelmed and de facto monopoly by North on news flow;
- Inequality in information resources—technological and channels;
- De facto hegemony and will to dominate—indifference to developing countries, information as a commodity, and capitalistic use of information;
- Lack of information on developing countries—only report the unfavorable;
- Survival of the colonial era—political, economic, and cultural colonialism;
- Alienating influence in the economic, social, and cultural spheres—propaganda, control of channels, advertising, opposition to social evolution, and cultural domination; and
- Messages ill-suited to areas in which they are disseminated—biased and irrelevant.

The continued pressing of these complaints within Unesco, along with a controversy regarding the licensing of journalists precipitated the U.S. and British withdrawal from Unesco in 1984 and 1985 respectively. Other activities sponsored by Unesco are less controversial from the U. S. point of view and deal with publishing output, not mass communication. These are UBC (Universal Bibliographic Control), UAP (Universal Availability of Publications), and UNISIST (Intergovernmental Conference for the Establishment of a World Science Information System).

UBC (Universal Bibliographic Control).   The purpose of UBC is:

> to make universally and promptly available, in a form which is internationally acceptable, basic bibliographic data on all publications issued in all countries. . . . The concept of UBC presupposes the creation of a network made up of component national parts, each of which covers a wide range of publishing and library activities, all integrated at the international level to form the total system. (Anderson, 1974, p. 11)

The reason for the program is that in many countries libraries were seen as static repositories primarily for the use of elites. Libraries have now come to be seen as a vital tool in economic and cultural development by providing an institution where literacy can be promoted and increased, and where economic and political information may be made available to the populace at large. The system's requirements mandate that each country ensure that it is *possible* to make a bibliographic record for each new publication issued in that country and that a national bibliographic agency will *exist* that will establish such records without delay, publish those records in a regularly published national bibliography, produce and distribute these records in a standard format for use around the country, receive and distribute bibliographic records similarly produced in other countries, and, as soon as possible, create a retrospective national bibliography of publications previously issued in that country (Anderson, 1974, p. 11). This system should also apply to serial publications and, in some cases, to their contents.

This goal of universal bibliographic control is the establishment of power over the world's publishing output. The power is the ability to find information about virtually any topic no matter where in the world a person happens to be or where the information is located. The UAP (Universal Availability of Publications) program attempts to make the information found available to the seeker.

UAP (Universal Availability of Publications).   The rationale for UAP was put forth in 1973 by the Committee on International Lending and Union Catalogues (quoted by Line and Vickers, 1983, pp. 14–15):

(i) That efforts to establish Universal Bibliographic Control will increase the demand on interlending services;

(ii) That in consequence the drive for UBC should be linked with a drive to improve international interlibrary services;

(iii) That in particular each country should aim to have, besides a national bibliography, a national centre or centres which will arrange to provide to other national centres on request a loan or a photocopy of any item published in the country.

The same committee a year later issued another statement, putting forth the fundamental principle of UAP (Line and Vickers 1983, p. 15, quoting Committee on International Lending and Union Catalogues):

As an essential element in any programme of national and international planning, and as a natural concomitant of Universal Bibliographic Control, efforts should be made both within each country and between countries to improve access to publications, by increasing the availability and speed of interlending services and by developing simple and efficient procedures. The ultimate aim should be to ensure that all individuals throughout the world should be able to obtain for personal use any publication, wherever or whenever published, either in original or in copy.

From this activity of international cooperation and the development of a global information system that relies on national bibliographic agencies, there arose several other specialized global information systems that complement both UBC and UAP. The most ambitious and important of these was UNISIST.

UNISIST (Intergovernmental Conference for the Establishment of a World Science Information System). UNISIST, which is not a very recent idea, became an official program of Unesco when Unesco and the International Council of Scientific Unions got together to form the system. UNISIST's fundamental approach can be described by Resolution 8 of UNISIST (Intergovernmental Conference for the Establishment of a World Science Information System, 1971, p. 17):

[The Intergovernmental Conference for the Establishment of a World Science Information System (UNISIST)] . . . acknowledges that, from the outset, UNISIST must ensure that adequate information for the utilization of accumulated knowledge in science and technology is available and fully accessible and that UNISIST must be based on voluntary co-operation between existing and future autonomous national, regional and international scientific and technical information services and systems, whether public or private, and that in this respect, the special needs of individual States must be taken into account, in particular those of developing countries; that initially, UNISIST must embrace the fundamental sciences, the applied sciences broadly conceived, technology, and subsequently, be extended to the social sciences and humanities as soon as possible; that such a World System must be multilingual, and must allow the use of national languages.

It is easy to see that if UNISIST were developed to the extent proposed, then UBC and UAP would be subsumed within it. As a matter of fact, ISDS (International Serials Data System) was implemented under the auspices of UNISIST and has "thus become the effective machinery for organizing one segment of the UBC system" (Anderson, 1974, p. 27). Other specialized systems and programs, such as AGRIS (for agricultural information), INIS (for atomic energy information), and MEDLARS (for medical information), have been started as a result of the broad concepts developed within UBC, UAP, and UNISIST.

Other Multinational Information Arrangements

Along with and subsequent to these global systems sponsored by Unesco, several networks and organizations have arisen in order to carry out cooperative ventures for information access and availability on a *regional* basis.

Europe.   In Europe several specialized databases and information systems have arisen, most of them sponsored or aided by the existence of the European Community or EUSIDIC (European Association of Information Services). Burkett (1983) describes many of these databases and networks, including Euronet/DIANE, the telecommunications network for Europe. The members of the DIANE network and of EUSIDIC have already developed a useful information system for their part of the world (*Information Policy for the 1980s*, 1979). This has been accomplished in part because of the developed state of the member countries' economies and the reliable telecommunications networks that exist in Europe.

Mahon (1989), who reviews the most recent developments in European information policy making, cites four trends affecting information policy in Europe. These are:

- Increase of accountability and use of the private sector to carry out tasks done by government;
- PTT (Postal, Telephone, Telegraph) deregulation undertaken so that there would be competition in the supply of information and related services;
- The gradual realization by publishers of the potential of electronic information; this led to pressure on the governments to make policy concerning what information is owned by the government and what information is open for exploitation by these publishers; and

- The EEC (European Economic Commission) is no longer seen as the coordinator of the information field in contemporary Europe.

Despite these trends, or possibly because of them, Mahon (1989, p. 86) worries that "since there is not evidence at either a national or European level of an overall examination of potential policy issues, no single body will be in a position to provide reasoned input to the policy decisions." No leadership seems to be forthcoming from countries on the perimeter of the EEC, namely Switzerland, Austria, Eastern Europe, and the Nordic countries. In these countries Mahon (p. 69) reports "information policy was mostly conceived to ensure an unburdened access to the information stores of other nations and also to provide some economies of scale for the preservation of information products in 'minority' languages."

Muslim World.   A different situation exists in the Muslim world, where information systems are in planning stages. Besides the problems of lower economic development and lack of country-wide telecommunications services, these systems are not yet as developed as those in Europe. Indeed, one primary concern seems to be the ironic state of affairs of wanting to develop information systems based on a Westernized model, while at the same time rejecting many of the values that gave rise to the model to be appropriated. Sardar (1988, p. 161), for example, states:

> Since much information that is generated in the modern world has little relevance to Muslim societies, Muslim countries would have to focus on generating their own information and knowledge base. This becomes even more urgent given that information is rapidly becoming a major source of power, and access to information would shape the destiny of states in the future. As such it is necessary for Muslim countries to generate their own information; that is, Muslim countries must develop self-sufficiency in local, relevant and significant R and D capabilities as well as domestic technological self-reliance. The experience of three decades of conventional strategies of development has shown that reliance on external sources leads to a particularly ugly form of dependency and ushers in a new form of colonialism.

This paradox of wanting to accept certain aspects of Westernization while rejecting the dominance that comes with that acceptance is not limited to Muslim societies (Von Laue, 1987). While the only model for a developed national information system in the Muslim world is Malaysia (Mohamed, 1988), there are several plans for developing similar systems, for example, in the Gulf Region (Ashoor, 1988), for the Islamic world in general by forming Islamnet (Kamaruddin, 1988), or by building on existing cooperative Islamic organizations such as the Arab League Educational Scientific and Cultural Organization (ALESCO) or the Gulf Cooperation Council (Rahman, 1988).

Southeast Asia. In Southeast Asia Brunei, Indonesia, Malaysia, Philippines, Singapore, and Thailand have undertaken information networking activities through ASEAN (Association of Southeast Asian Nations). Their problem is similar to the Muslim nations. Since North America, Europe, and the USSR produced 67.1% of the world's books in 1987 (*Unesco Statistical Yearbook*, 1989, Sec. 6, fig. 6), and the United States and Europe alone produced 89% of the databases in 1989 (*Computer-Readable Databases*, 1989, p. xiii), there is a threatening dependence by these countries on the developed nations. One of ASEAN's fears is that their needs will not be met by the information superpowers (Lim, 1989).

South America. In South America a regional network has not yet developed, although networking activities in several countries have been initiated (*The Transfer of Scholarly, Scientific and Technical Information between North and South America*, 1986).

Africa. In Africa PADIS (Pan African Documentation and Information System) is "a regional information system and network at the service of development" (Abate, 1989, p. 81). PADIS arose after several international systems and programs had already been implemented in Africa with varying degrees of success. These programs, however, could not capture enough of the available literature and, therefore, did not achieve much progress (Abate 1989). Two lessons learned in establishing PADIS were:

- Overlap between systems and sub-systems in [sic] unavoidable. This can be resolved only as the network matures; and
- Partners in a network are invariably at different stages of development. Therefore, it should also be the aim of networking programmes/projects to assist those at lower levels of development (Abate, p. 82).

Soviet-Dominated Communist Countries. In the formerly Soviet dominated communist countries, an agreement among the members of the Council for Mutual Economic Assistance was signed on February 27, 1969 for the establishment of an International Center of Scientific and Technical Information (Mezhdunarodnyi tsentr nauchnoi i tekhnicheskoi informatsii) [hereafter MTSNTI]. On the basis of this center, located in Moscow, the member countries (Bulgaria, Hungary, East Germany, Cuba, Mongolia, Poland, USSR, Czechoslovakia, Romania, and Vietnam) created an International System of Scientific and Technical Information (Mezhdunarodnaia sistema nauchnoi i tekhnicheskoi informatsii) [hereafter MSNTI]. The center was to complement already existing Soviet indexing and abstracting agencies, such as VINITI (Vsesoiuznyi institut nauchnoi i tekhnicheskoi informatsii [All-Union Institute of Scientific and Technical Information]) (Mikhailov, Chernyi, and Giliarevskii 1984), by covering to the greatest extent possible the literature published in the member states, by establishing standardization and compatibility of information transfer mechanisms, and by automating information processes. The system had two major goals: (1) to raise the level of satisfying information needs among the participating countries; and (2) to gradually improve the several national information systems in order to raise the level of the cooperative international system (Gavrilova, Kuskov, & Tyshkevich, 1981, p. 22). In spite of the acute awareness of the information explosion and the importance of information systems to deal with the explosion (Kashlev, 1988; Mikhailov, Chernyi, and Giliarevskii, 1984, Chapter 1), the Soviets and their allies could not make the system work as they had envisioned it. One of the primary causes of this failure was that poor telecommunications systems within these countries prevented the reliable transmission of data over the conventional telephone networks.

In spite of these shortcomings, the Soviet-dominated pattern (at least prior to the Fall of 1989) had four goals for their style of information society:

- To attain real gains in productivity and to modernize the industrial base;
- To improve the economic planning and control mechanism;
- To support both military and internal security needs; and
- To present the image of a progressive society both to the people of the USSR and to the outside world (Goodman 1988, pp. 13–14).

Similar expositions of the Soviet vision or its derivatives are given in Judy and Clough (1989) and Richards (1986).

For all these countries, developed and developing, Goodman (1988) sees some common ground in the problems faced. He thinks that "the most important common thread that emerges from our models of both Western- and Soviet- style information societies is the need and desire to improve control over increased complexity and reduced time scales" (p. 17).

This brief coverage of some of the multinational activities aimed at providing information interchange indicates the complexity of the information policy maker's task. Because many programs, networks, and arrangements (some of which already overlap) already exist, any information policy is bound to require an immense amount of prior planning just to see what effects it might have on already existing structures and arrangements. It would be a mistake if policy makers believed this to be the extent of the complexity with which they must deal.

ETHICAL CONCERNS

Perhaps the most important factors affecting information policy formation, which are often relegated to the end of the discussion, are the values and ethical concerns affecting the assumptions of information policies. Ethical concerns are often secondary to or implicit in the acquisition of knowledge and the making of viable policies. Policy makers generally do not start with the ethical implications, often because they may not understand what these implications are until they have examined the area over which decisions must be made. Placement of the discussion of ethics at this point in the chapter is based on that rationale. However, that discussion is, or should be, the primary consideration because information policies are inescapably value-laden. Who has access to what information? Who can control the dissemination of information? What importance do we attribute to information and its processing? These are all questions of value. How these questions of value are resolved is an important task of information policy formulation, review, and implementation.

Several ethical and justice issues have already been identified with international information transfer. Two of the most visible of these are personal data protection and the disparity between the information "haves and have nots." Again, these are presented primarily as examples of one of the many, albeit most important, considerations affecting information policy making.

Privacy

The right to privacy, or the right to be left alone, has gradually gained grudging acceptance as a human right. Eres (1989, pp. 7–8), citing others, lists several principles that regularly appear in the privacy statutes of many nations. These principles are:

- Openness (no secret personal data recordkeeping systems);
- Individual access (the right to know what data are on record and how they are being used);
- Individual participation (the right to amend or correct personal data about oneself);
- Collection (limits on the types of data that can be collected and the use to which they can be put);
- Use (right to limit the use of one's personal data to the use for which they were collected);
- Disclosure (limits on the external disclosure of personal data that have been collected by an organization);
- Information management (requirements for the implementation of data management policies by recordkeeping organizations); and
- Accountability (accountability of recordkeeping organizations for their operations regarding the data).

The extent to which any of these principles are included in a given nation's privacy statutes depends on cultural, political and economic factors particular to that country. It goes without saying, of course, that since nations' privacy laws differ, anyone dealing in international information flows will have to contend with conflicting requirements for privacy and data protection between countries and with laws that may conflict with whatever goal is targeted for achievement by another nation or by multinational corporations. Salvaggio (1989), who takes this one step further, wonders whether personal privacy will be possible any longer in a society where information technology is becoming ever more capable of storing, retrieving, and disseminating personal data.

Information Haves and Have Nots

The MacBride Report (MacBride, 1980) was the result of a two-year study of the current problems facing world communication. This included, but was not limited to, issues of justice and equality in what was perceived to be an unjust and unequal environment. The report challenged the information-rich nations to acknowledge and to rectify the gross inequalities and injustices that resulted from the developed world's hegemony over information technology, information products of all kinds, and robust and resilient information infrastructures.

After a thorough examination of "Communication and Society," "Communication Today," "Problems and Issues of Common Concern," "The Institutional and Professional Framework," and a visionary "Communication Tomorrow," the report concluded with 82 recommendations that "indicate the importance and scale of the tasks which face every country in the field of information and communication, as well as their international dimensions which pose a formidable challenge to the community of nations" (MacBride, 1980, p. 272).

Rather than attempt to summarize the entire report, this chapter will mention three foci of concern that the report itself deemed "particular challenges." These are paper, tariff structures, and the electro-magnetic spectrum.

Three MacBride Report Challenges

Paper.   With a worldwide shortage of paper, and escalating prices for the available supply, many Third World countries "impose crushing burdens upon struggling newspapers, periodicals, and the publication industry, above all in the developing countries" (p. 257). The report gave various alternatives for supplying additional paper substitutes and recommendations for increased use of recycled paper.

Tariff structures.   "Tariffs for news transmission, telecommunications rates and air mail charges for the dissemination of news, transport of newspapers, periodicals, books and audiovisual materials are one of the main obstacles to a free and balanced flow of information" (p. 257). Above all, the report urged governments not to look upon these revenues primarily as potential profit but instead they should be part of larger national goals. The report also urged international bodies to stop or curtail the imposition of tariff structures that especially hurt small and peripheral users.

Electro-magnetic spectrum The spectrum and the geostationary orbit, "both finite natural resources, should be more equitably shared as the common property of mankind" (p. 258). The example of the Kingdom of Tonga claiming several telecommunication satellite parking slots in the geostationary orbit was previously discussed.

In each of these cases the disparity between information-rich nations—such as the United States, the EEC, and Japan—and information poor nations is clear. Any policy maker must take these types of disparities into consideration in policy making. If such issues are ignored, the current situation is perpetuated and could become more acute, resulting in policies with unintended and possibly harmful effects.

SUMMARY

This chapter has turned to the international context of information policy making in order to specify the types of considerations that must be included in an evaluation of information policy. The solving of these problems cannot be accomplished using only one or two of the existing specialized fields of study (political science, law, economics, or information science), nor can they be solved without making decisions about value. Resolution of these policy-making problems can involve virtually any disciplinary knowledge that now exists.

Even if all the relevant knowledge needed for evaluation could be identified, policy making may not necessarily proceed as designed. Since the real world contains unforeseen events, there should be no illusions about the perfectability of information or any other policy making. The concern of this book is with the current state of information policy and the present ability to design and to evaluate it, however limited that ability may be. The aim is to inject an evaluative framework into the evaluative process that will make that process more comprehensive and more consciously ethical. The coming chapters narrow the focus of this discussion to scientific and technical information policy in general and the SATCOM report in particular. The intention is not simply to gape in wonder at the complexity of designing and evaluating information policy, but to begin to ascertain how adequate our existing knowledge is for this task, specifically within the realm of scientific and technical information policy.

CHAPTER 3

SCIENTIFIC AND TECHNICAL INFORMATION (STI) POLICY AND THE FUNDAMENTAL PROBLEM OF INFORMATION POLICY EVALUATION

The use of supercomputers in science and technology has enabled humans to predict with fairly good accuracy a variety of naturally occurring phenomena. For example, Markoff (1991, p. C1) describes supercomputer applications in environmental science:

> Supercomputers can study the weather through the use of mathematical models that account for dozens of factors, like atmospheric pressure, temperature, regional topography and wind direction. By breaking up a large region into many small areas, highly accurate forecasts can be developed.

His ascription of human capabilities to supercomputers (studying the weather) notwithstanding, Markoff has described one of the more impressive applications of information technology. What he understandably omits from his description, however, is the process by which the meteorologists using these supercomputers have acquired the data that constitute the "dozens of factors" that are the raw material processed by the supercomputer. The report of the predicted effects was generated by processing raw data acquired by satellite technology, weather instrumentation, and a host of other devices and methods. The meteorologist using the supercomputer had managed to retrieve these data and, with a sound theory about how the variables interacted with one another, produced a prediction of what the future weather would be under given conditions.

A scientific and technical information policy exists in the organization where the meteorologist works. This policy enables the meteorologist not only to acquire the information needed but also to dictate how the information produced from the modeling experiments will be disseminated to the public. Questions about the pricing of this information, the specific means by which the information is generated, how much would be paid for the raw data, who would have access to this raw data within the organization, the means by which and the format in which this information would be disseminated to others are all part of the information policy of the organization. The success with which weather patterns are predicted accurately depends on a sound meteorological theory, reliable data, human meteorological expertise, the appropriate technology to process the dynamic weather model, as well as the information policy of other organizations with which the weather service interacts. For if a research institute that gathered wind speed data would not release it to the public, or if it did release the data but in a form unamenable to easy manipulation, the scientist who needs this variable as part of the meteorological model would have to spend extra resources in making this data amenable to the modeling program. The other alternatives would be to gather this data using the organization's resources, or else do without it. The result of this latter course of action would be to render the model ineffective and unreliable.

One can imagine the resourcefulness of the meteorologists and the information specialists who must not only determine what information is needed to create a reliable model, but once this determination is made, to ascertain the source of data to fulfill this need. Many avenues will be explored to achieve this goal: results of other models and theoretical documents on the subject, reports from Federal laboratories that produce such data, personal communications with other researchers involved in the field, and attendance at conferences. It may also involve translation services to obtain information from abroad, from other nations' researchers who are working on the same problem. This system of scientific communication is the area over which STI policy seeks control. (For background reading on the topic see, for example, Garvey, 1979, and Allen, 1985.)

In spite of the myriad suggestions advanced over the years to make scientific information easier to access both on an institutional level and a personal level, no robust theory has emerged that will predict with any reliability what actions must be taken in order to promote the most efficient and effective use of scientific and technical information on a national scale. The construction of such a theory may not, in fact, be possible; if there exists such a theory, it has not been well articulated nor made the cornerstone of any national policy on scientific and technical information. This is not to say that we have failed to promote the efficient and effective societal use of scientific and technical information. What it means is that in spite of the many suggestions about improving the scientific and technical information system in this country, these suggestions have yet to inhere in any coherent national policy.

THE LACK OF A NATIONAL SCIENTIFIC AND TECHNICAL INFORMATION POLICY

Unfortunately, there is no one document to which one can point and identify as the United States' national Scientific and Technical Information Policy. As Gould (1986, p. 61) states,

> National scientific and technical information policy, like many broad areas of government concern, is not anywhere articulated in a comprehensive form. Agencies that sponsor basic and applied research in support of broad mission needs—such as the Departments of Defense and Energy and the National Aeronautics and Space Administration (NASA)—each have their own policies and practices for dissemination and access to information produced. Those seeking to understand scientific and technical information policy must derive it from numerous statutes, legislative histories, regulations, and executive branch directives.

While the goal of such policies is often assumed to be the generation, management, and disposition of scientific information, there have been many arguments over the ways in which this should be done. In summarizing more than 20 years of studies and proposals, Chartrand and Chalk (1975, p. 60) declared:

> The importance of scientific and technical information as a *national resource* has only recently emerged. It is difficult for non-technical personnel to understand that one of the most significant products of the sizable Federally-funded research and development programs is *information*. These voluminous data and accompanying narrative insights presented in varying forms, can be fully utilized only through a strategy for their generation, management and disposition. This strategy is not apparent in today's STINFO operation. To fulfill these interlocking tasks implies the presence of an assertive leadership, aware of the complex needs of this post-industrial, civilian-oriented society and dedicated to ensuring that that combination of authority, responsibility, and flexibility necessary to carry out this priority mission is functioning. Today, there is a widely alleged lack of such leadership, suggesting strongly that 'management' of a precious resource is fragmentary and questionable at best.

This document is representative because it not only summarizes previous studies of the STI situation in the United States, but it also makes many of the same assumptions regarding the solution of STI problems. The most telling assumption, of course, is that the problem with STI policy is not with any lack of knowledge about the actual management of information flows, but the lack of leadership in using knowledge that we already possess. The means of controlling, disseminating, and processing information are all assumed to be known. Unfortunately, there is little evidence to suggest that we are knowledgeable about these means on a national scale. Many whose optimism in this regard is unwarranted seem to be saying that if only we could muster up the leadership to manage and put into effect all our knowledge, we could solve this problem.

One reason for laying the blame for the failure of previous STI policy on national political leadership is that evaluations of STI policy suffer from a lack of in-depth analysis. The literature is either derivative, repeating previously suggested solutions, or is speculative, suggesting new ways for effective political implementation of proposed policies. The issue of the soundness of individual recommendations and their theoretical justification has been raised only to a limited degree.

THEORETICAL SOUNDNESS OF STINFO POLICY RECOMMENDATIONS

Bishop and Fellows (1989, p. 33), in reviewing several information policy proposals over the past 30 years, found that many recommendations were made without any suggestion of their probable impact. This lack of a predictive component in scientific and technical information policy documents is indicative of an absence of sound knowledge about the phenomena over which the policy intends to exert control. Policy failure or lack of impact has therefore been attributed to the vagueness of the recommendations proposed, to the comprehensiveness of the solution proposed, or to the inability of policy proposers to communicate their needs adequately to policy makers. These deficiencies are not unique to scientific and technical information policy; they are similar to educational policy or health care policy, for example. Political and organizational deficiencies are only part of the problem, however. The argument made here is that our knowledge of information flows is inadequate for the policies we wish to formulate. In many respects we have condensed microinformatics, the study of information phenomena within specific institutions or organizations, with macroinformatics, the study of information phenomena in the large.

Our discussions of scientific and technical information policy reflect this bias. For example, those advocating a centralized coordinating body for information policy assume, do they not, that this central body will have the power to affect information flows in line with approved policies. In order to do this, in much the same way that the Federal Reserve Board controls the flow of money throughout the economy, the assumption is made that we have the theoretical knowledge that would suggest the wisdom of taking specific courses of action. But we do not possess this knowledge, at least not in any codified, systematic form.

In a similar manner those advocating a decentralized approach to information policy still argue for some central national policy that at a minimum specifies that information produced by the Federal government be made available to the public for their use. This approach also assumes some operating principle that assures the use of information by specific agencies, organizations, or by individual researchers if only information were made available. The purpose of this chapter is to explore these approaches in more depth. To do this the chapter will first describe what is meant by scientific and technical information policy and why it is important. Following this an overview of STI policy in the United States points out the goals that policy has sought and the successes and failures of its programs.

WHAT IS SCIENTIFIC AND TECHNICAL INFORMATION?

One would assume that prior to suggesting a policy concerning an area, the boundaries of that area would be delineated. This has not generally been the case with scientific and technical information policy. Most Western writers cast a wide net when discussing scientific and technical information. This includes, but is not limited to, articles appearing in scholarly journals, technical reports, patent information, books on technical or scientific subjects, sound recordings, videodiscs, and other similar media that describe scientific processes or explain engineering principles. In addition, raw data from weather satellites, unprocessed data on geological formations, and data from space probes are included under the rubric of scientific information. It includes oral presentations at scholarly meetings, scientific and technical devices in which information inheres, technical advertising material, and personal contacts between scientists during which science or technology is the subject of conversation. It also encompasses information about devices and techniques that encrypt and transmit other information. In some extreme cases, information about military forces or the operation and viability of commercial enterprises would also be included.

For Mikhailov, Chernyi, and Giliarevskii (1984, p. 69) scientific information "may be visual, aural, and tactile (palpable)" and is distinguished from data. For them (Ibid.) *data* means "information received in the process of emotional cognition that has not yet been subjected to the processing and generalization of abstract/logical thought." To explore the distinction between data, information, and knowledge would require an extended study in and of itself. Such is not the purpose here. The point to be made, for this investigation, is that policy makers and analysts have not been rigorous with regard to what it is they are making policy about. Such inattentiveness to the realm over which control is to be exerted dilutes the potential robustness of any policy.

This sloppiness with regard to the definition of scientific and technical information has come about from an ignorance of how to control this information for various purposes. On an institutional level, the task seems straightforward, and often is. Once we move beyond the individual organization, however, our knowledge about information and the effect of social and cultural, as well as technological, forces is sparse indeed. If we are not knowledgeable about the effects of our efforts to control information, effects that go beyond moving a copy of one document or computer program or dataset from one place to another, it is no wonder that there is no precise definition of information, except by Shannon and Weaver (1949) in their classic article about information. They focused on the accuracy of signals received through a telecommunication link. Their definition, although rigorous and unambiguous, does little to help the policy analyst define the policy area over which control is to be exerted.

The definition of information is important. However, the inability to define it should not prevent the examination of the commonly held view of scientific and technical information—how such information is created, represented, stored, disseminated, and used. The following discussion adopts a microinformatic perspective; that is, it represents the point of view of an individual organization. The policy problem arises when more than one organization is involved in processing and using this information. This macroinformatic perspective will follow.

The Creation of Scientific and Technical Information

My 12-year-old daughter recently performed an experiment for her science class. She had become intrigued with changeable figures, those two-dimensional representations of objects that can appear in two different ways to a human being. An example is the picture that can be perceived as either an hour glass or two faces confronting one another. She formulated a hypothesis regarding the differing ability of males and females to perceive the "other" object in a changeable figure. She then tested her hypothesis and with my help applied a statistical test to determine if the average time required by the two groups to perceive the second figure differed significantly. (They did not.) Is her completed class report scientific information? Well, yes and no. It followed a widely accepted scientific method to test her hypothesis. It has all the trappings of a scientific study. The results were produced in written form. In spite of meeting these criteria, however, I would doubt that most would accept this as scientific information, primarily because it had not been judged by "peers." Her peers certainly judged it highly. But peer review in this sense means review by others who have the credentials to judge such experiments and who have given their approval to publish her results in a scientific journal.

This example suggests that scientific and technical information is not simply judged to be so by virtue of following agreed-upon procedures for its production. It has a social and cultural element to it. Elements that were included in my daughter's experiment (creativity, hypothesis testing, sampling, and statistical techniques) are not sufficient to call the results of her study scientific information. Yet, we label wind speed data collected by scientific instrumentation, but not yet interpreted by a human being except to give it in a standardized measure of wind speed, as scientific information.

Scientific and technical information is defined according to criteria developed and acknowledged within the scientific community. The creation of scientific and technical information occurs when scientists and others legitimated by society for performing scientific and technical work, carry out investigations to test their hunches in their field. To do this they may use scientific instrumentation to gather data about the area in question. Reports of their investigations are scientific and technical information. The data gathered during this process are also called scientific and technical information.

The Representation of Scientific and Technical Information

Scientific and technical information can be represented in the traditional form of printed documents and brochures and also on various other types of media, such as computer tape, computer memory devices, videocassettes, and audiocassettes. The scientific article is usually thought of as the prototypical format in which scientific and technical information appears. To assume that this is the predominant form of STI representation, however, would be in error. Technical reports produced within both private and public laboratories, advertising brochures depicting equipment and instrumentation along with printed technical specifications relating to the item portrayed, printed and machine-readable computer programs, preprints of articles that are scheduled to appear in scientific journals, and reviews of the literature of a specific topic in article or book form are all widely used ways of representing scientific information. In addition, maps, charts, diagrams, patent information, photographs, slides, microforms, and other similar media are also common forms of representation.

The Storage of Scientific and Technical Information

Scientific and technical information within the possession of an organization does not simply lie around on tables waiting for the members of that organization to use it. It is either stored by an individual researcher in a place to which that researcher may restrict access, such as an office or private file cabinet, or by the organization itself in some kind of central repository or library. In either case there is usually some logical method of arranging the information so that it may be retrieved and examined when needed. The storage and retrieval of information of all kinds is a well developed discipline in its own right and is central to scientific and technical information policy (Mikhailov, Chernyi, and Giliarevskii, 1984, Chapters 7-9). The privilege of free and unimpeded access to these stores of information by members or owners of an organization also figures prominently in an organization's information policy (Stevenson, 1980).

The Dissemination and Use of Scientific and Technical Information

To raise the question about how scientific and technical information produced by an organization is to be disseminated is also to ask why it should be disseminated. In answering this question one would include the reasons why such information was created in the first place and who else besides the creator would want that information. In some cases the information was produced as the final step of basic scientific investigation. To disseminate this kind of information is to claim and to establish priority for a specific scientific discovery and to share this discovery with other researchers in one's field. Here full public disclosure is sought and encouraged.

In other cases the information to be disseminated was the result of some type of mission-oriented research. Questions relating to the kind of mission, the dangers of letting rivals know such information for fear they will advance the discovery further before the original investigator does and accomplish the mission that person set out to complete, the prior restraint placed on the investigator by the agency funding the research, the implications for national security, and cost all affect the decision of dissemination. In these cases the survival of the organization or its sponsor may be at stake.

The assumption about dissemination, of course, is that the information will be used. The assumption further made about information use, especially scientific and technical information use, is that it has effects beyond the scientific establishment itself. For example,

> The United States must make better use of its scientific and technical information (STI) resources, if it wishes to be competitive in world markets and maintain its leadership. STI is an essential ingredient of the innovation process—from education and research to product development and manufacturing. . . . Many issues of our time—health, energy, transportation, and climate change—require STI to understand the nature and complexities of the problem and to identify and assess possible solutions. STI is important not only to scientists and engineers but to political, business, and other leaders who must make decisions related to science and technology, and to the citizens who must live with the consequences of these decisions. (U.S. Congress. Office of Technology Assessment, 1990, p. 1)

Furthermore, effective use of STI brings many benefits and savings for the society at large:

- Time saved in locating other researchers doing related work;
- Time and money saved in minimizing duplication of research effort;
- New insights or breakthroughs resulting from more complete awareness of related research;
- New information not available elsewhere;
- Better understanding of relevant Federal R&D directions; and
- Time and money saved in writing research reports, papers, and articles (Ibid., p. 12).

Because of these kinds of implications for the society at large, one would think that governments would have taken an active interest in STI policy. Such has not been the case, as years of various STI policy proposals have demonstrated (Bishop and Fellows, 1989). Prior to the investigation of what efforts have been made in the United States in this regard and an assessment of these efforts, it is first necessary to examine STI from a macroinformatic perspective.

THE MACROINFORMATIC PERSPECTIVE

Macroinformatics looks at information phenomena in the large, not from the point of view of individual institutions. Macroinformatics is not a developed discipline, nor even yet a field of study. Use of the conception of macroinformatics at this point is simply a heuristic tool to help us see both the complexity faced by STI policy and the lack of a theoretical base upon which to base national-level policy.

One problem confronting a macroinformatic perspective is that of the relation of the parts to the whole. Langlois (1983) describes two poles of the part/whole debate:

> The conventional wisdom runs something like this. There is a doctrine in the social sciences called methodological individualism. It is a form of the analytic or reductionist method, and it therefore holds that knowing the properties of the parts—the individuals in society—is fully sufficient for grasping all there is to know about the whole—society. In other words, the properties of the whole can be deduced from the properties of the parts; or to put it in more familiar (and more naive) terms, the whole is just the aggregate or sum of the parts. This view is to be contrasted with methodological holism (or sometimes collectivism), which insists that wholes possess "emergent" properties that cannot be derived from the properties of constituent parts. To the holist, the whole is greater than the sum of its parts. (pp. 582–583)

Langlois (1983) proceeds to support a stance that combines the best of these approaches, and which at its root is what "sophisticated methodological individualists believed all along" (p. 586). "The behavior of the whole cannot be understood without knowledge of the relations among the parts" (p. 585).

This framing of the issue in these terms has implications for the formulation, implementation, and evaluation of STI policy. Although the present mental model for a national system of scientific and technical information is analogous to the behavior of an individual organization's system of scientific and technical information, the analogy is erroneous on at least seven counts. First, the scale of an individual organization in no way approaches the scale of a national system. Although policy analysts realize this, they still retain the single-organization model for the entire nation.

Second, analysis of the behavior of an individual organization consists of an examination of the behavior of people within that organization. The organizational behavior observed is, therefore, an aggregate behavior pattern of those individuals. The national behavior is an aggregate of these aggregates. Can we assume that an analogy can hold when one element of the analogy is based on individuals and the other on aggregates?

Third, and this is a corollary to the second, different organizations have different goals, even within science and technology. An individual organization's behavior consists of that organization's goals, the leadership within it, the personality of the individual humans working there, the resources at its disposal, the larger environment in which it works (academic, governmental, or corporate business), and the branch of science or technology in which the individual researchers are engaged. An organization involved with basic science has a different time line and a different approach to problems than does a mission oriented organization.

Fourth, STI is an important element common to all these organizations. It is not always, however, a controlling element. In the present macroinformatic viewpoint, an individual organization's behavior is not only aggregated, but its goals are also reduced, primarily to information seeking and processing. This cybernetic view of society has received tempered criticism elsewhere (Stanley, 1978, Chapter 6).

Fifth, it does not include any definable structure by which all these aggregates (organizations) are connected. If we do not have a model of how they interact, then any policy made to alter or control their behavior comprises guesswork.

Sixth, it often, but not always, ignores the value and quality of information. It assumes that, as Langlois comments (1983, pp. 586–587), information flow is like an oil flow directed to society's storage tank and results in a tank of knowledge. This task, however, contains a heterogeneous, not a homogeneous, mass, varying in reliability, verifiability, structure, and relevance. The contents possess different values. Although we know a great deal about how many scientists and engineers use information, and how they choose the information they want from a plethora of information around them, these insights have not been incorporated into macro policies. On the macro level, all STI does not have the same value. The problem is a difficult one, but the primary goal is more than information access and dissemination. Access and dissemination, though, are *sine qua non* for the subsequent task of filtering and evaluation of the information to be used.

Finally, existing models (if that is what they can be called) are basically static ones of information use. This shortcoming has resulted from an ignorance of the dynamics of how information is used on a national scale versus how it is used within organizations.

The macroinformatic model is not a mechanistic one that will reveal the intricate workings of our information circulatory device. As shown in Chapters 1 and 2, it involves elements of political, technical, social, cultural, and ethical knowledge. With these caveats in mind, this chapter examines STI policy in the United States.

STI POLICY IN THE UNITED STATES

By its very definition STI policy implies governmental policy. Those who focus purely on the governmental side of the STI world are likely to miss at least half of the STI phenomena. Since 1980, for example, the private sector has contributed a larger share of monetary support for R&D than has the Federal government ($49.8 billion versus $44.5 billion) (Schact, 1985, p. 2, quoted in Ballard, 1987, p. 196). Although this is not a precise measure of STI effort, it indicates that a significant share of STI activities belongs to the private sector. Any national policy must incorporate the private sector.

Private Sector STI

In recent years a great deal of research has focused on the governmental provision of STI. This is understandable considering the fact that the Federal government supports almost half of R&D in the country, even if it spends only a small fraction of this amount on the dissemination of research results (McClure, 1989, p. 3). By and large the non-Federal contribution to U.S. STI is disseminated in several formal channels of communication. First, individual researchers both in academia and industry contribute to scholarly journals and make known the results of their research through this medium. Private databases provide access to this information, and private database vendors make these databases available to the public for a fee. Because of the large volume of research results in even one specialty, suggestions have been made over the years (Weinberg, 1963; *Scientific and Technical Communication*,

1969) to produce evaluative summaries of the literature in individual specialties on a regular basis. As discussed in Chapter 2, for information dissemination in general, some kind of filter is needed to evaluate the information produced as a result of R&D. These suggestions have not produced any great outpouring of annual reviews, although many fields have such reviews. The inertia in producing such documents is a result of the reward system in science. Individual researchers are rewarded for new discoveries and for establishing priority, not for literature evaluation. Any comprehensive STI policy must take this and other social phenomena into consideration. Some suggestions have been made for accomplishing this, including the provision of an incentive system within academic science whereby the authorship of such reviews would receive recognition and compensation equal to that of original research. The argument made is that the writer of a review also applies creativity to the authorship of such reviews—the creativity of synthesis. So far there has been little institutional change in this regard.

Academic researchers have also recently taken on some of the trappings of industrial researchers in that they apply for patents in conjunction with the college or university where they are employed. This patent information is accessible to other scientists.

Information dissemination of research results also occurs through informal channels of communication, such as telephone conversations between researchers, conferences, and site visits to other laboratories.

The non-Federal sector also consists of scientists and engineers in private industry. The result of their research is more often product- or mission-oriented. As opposed to the academic researcher, however, the industrial researcher not only disseminates the results of research through formal channels of scientific communication, such as journals and patents, but often withholds dissemination of research through trade secrets (Relyea, 1986). For many corporations, there has been a shift away from patenting to protection through trade secrets. Stevenson (1980) gives several reasons for this shift:

First, it is usually a good deal cheaper simply to keep a new technological development secret than to incur the substantial legal costs of acquiring—and defending—a patent. Smaller firms are at a particular disadvantage in this respect because of the enormous amounts of time and money typically consumed in patent litigation. They are often forced to yield to big companies in patent contests simply because of the ability of their larger competitors to outspend them. This is, in fact, one of the most serious defects in the patent system, particularly in light of the generally accepted proposition that smaller firms are responsible for a disproportionate share of technological innovations.

Second, not only is it generally less expensive to maintain the secrecy of a trade secret than to obtain effective patent protection, but the former is more likely to be successful in achieving the end sought. Historically, about half of all lawsuits brought to defend trade secrets are decided in favor of the plaintiff, whereas patent-infringement suits end in a favorable result for the patent owner only 20–30 percent of the time.

Third, the fact that patent applications are matters of public record also tends to lead innovators to rely, where possible, on secrecy. Not only do they lose any protection they might otherwise have had from the law of trade secrets if a patent is denied, but the publication of a patent that is granted frequently serves as an invitation to others in the field to "invent around" the patent.

Finally, not all innovations are patentable. A new development that does not meet the patent statute's rather rigorous (at least as interpreted by the courts) standards of novelty, utility, and nonobviousness may nevertheless be of substantial commercial value so long as it can be kept secret from the competition.

For all the apparent advantages of the law of trade secrets for the innovator, however, the extent to which it results in a net benefit to society is not altogether clear. (pp. 21–22)

This practice of withholding information for the purpose of further innovation is one that policy analysts will also have to contend with in formulating national STI policy. To date, however, the pressure instead has been put on the Federal government to disseminate its R&D results to the tax-paying public.

Public Sector STI

The Federal government has been involved in a formal way with scientific and technical information policy since 1790 when the first Patent Act was passed to protect the rights of inventors. It took another 80 years, however, before the Patent Office would have the means to create a system of patent information, with the publication of the *Official Gazette of the U.S. Patent Office*. This publication disseminated patent information to U.S. science and industry (*The Story of United States Patent and Trademark Office*, 1981, pp. 14–15). The story of the Federal government's involvement in scientific and technical information is available elsewhere (Adkinson, 1978; Bishop and Fellows, 1989; Doty and Erdelez, 1989; Hernon, 1989). This history has yet to produce a Federal STI system that functions well. Why is this so?

The Federal government produces STI in ways similar to the private sector. Its scientists publish in scholarly journals, issue technical reports, gather and make available scientific data, and communicate with other specialists in their field, both inside and outside of government. The Federal government also contributes monetary and other resources to the private sector in order to carry out research programs of particular interest to the government. In a manner similar to trade secrets, the Federal government does not make certain research available for fear that it may endanger national security or erode U.S. competitiveness.

In contrast with the private sector, however, which seems cognizant of the importance of STI for the survival of individual research organizations, the public sector has had problems in convincing Congress and the public of the importance of STI. STI has what has been called a secondary status among policy areas in the Federal government. Doty and Erdelez (1989, p. 70) believe this secondary status has come about for several reasons. First, "it is virtually impossible to give a simple, unitary definition of STI and to extricate STI from the policy instruments that affect it directly or indirectly. This situation also leads to fundamental conflict among stakeholders." Second,

> an understanding of information per se and its value, especially as a social good, is still difficult to achieve, despite the amount and depth of research done on these topics. For example, political and social theory has been unable to develop sufficiently useful models of cost-benefit analysis in order to give policy makers flexible, reliable methods by which to conceptualize and weigh various information policy alternatives. In addition, too little is known about the role of information in the innovation process. These conceptual problems lead directly to the general social inability to define, understand, measure, and appreciate the importance of Federal STI. (Ibid., p. 70)

Doty and Erdelez (1989) conclude by stating:

> STI is expected to serve too many masters, and this expectation, for many reasons, creates waste, inefficiency, polarities incapable of compromise, and policy fragmentation. In order for STI to be more fully utilized, it is imperative that scientific and technical information be recognized by all stake holders as an autonomous and legitimate policy field, for policy action and analysis. (p. 82)

Perhaps one reason why it is not recognized as "an autonomous and legitimate policy field" is that it lacks a theoretical base for predicting, within the limits of any social or cultural policy, the results of proposed policies.

This possibility, however, is not recognized by most studies on the subject. Instead, the major problem with Federal STI policy is the lack of coordination among the various agencies and governmental stakeholders. For example, a recent study (U.S. Congress. Office of Technology Assessment, 1990) claims that:

> Executive branch leadership is imperative because STI is generated by many Federal R&D agencies that must be coordinated if the government's STI efforts are to be successful. Agencies have set up a variety of ad hoc coordinating mechanisms for specific aspects of STI; but an overall, integrated approach is lacking. One of these existing committees could be expanded and chartered to serve a broader purpose. Alternatively, a new high-level interagency committee on STI could be established, with representatives from the R&D programs that generate STI, the agency data centers and technical document distribution offices, and governmentwide dissemination agencies such as the

Government Printing Office (GPO) and National Technical Information Service (NTIS). (p. 1)

The assumptions behind these administrative and organizational recommendations are questionable. On what basis can it be assumed that once the administrative and organizational functions are straightened out and the responsible bodies have the power and authority to implement changes, there will be a body of reliable theoretical knowledge available to which policy makers may turn to make the necessary decisions?

This should not be mistaken as an argument against the current research in STI, or any other information policy, concerning the political and organizational factors involved in effecting desired policy. A caveat is that the STI policy problem cannot be solved by focusing solely on organizational and political factors. Fascination with these factors has obscured the dearth of knowledge policy makers actually possess for making these decisions. Indeed, after examining the major U.S. STI policy studies, Bishop and Fellows (1989) remarked:

> A discussion of the probable consequences of implementing recommendations is virtually nonexistent in the STI policy studies examined. Rarely are the possible interactive effects of recommendations with existing policy instruments and current programs explored. Immediate and long-range economic, social, and political consequences are all given short shrift. It would seem that the exploration of the consequences of policy actions would be of interest and of value to policy makers trying to reach decisions. In general, however, the studies devote almost all of their attention to identifying STI policy problems and issues, and to developing concepts for solutions. (p. 33)

The solutions offered in previous policy studies are based on partial knowledge, political and organizational knowledge of the phenomena that policy makers wish to control. The solutions do not incorporate knowledge of the entire range of problems and areas that they need to address.

SUMMARY

This chapter suggests that policy makers do not have the knowledge available to them to formulate STI policies which are totally effective. This lack of focus on current knowledge, the lack of a broad theoretical base, and the diffusion of information among various disciplines prevent them from designing proactive and forsighted policies. The first step in formulating the necessary theory is to determine what existing disciplinary knowledge will be useful in information policy formulation and review. Part II of this book addresses this problem.

PART II

INFORMATION POLICY LIMITS AND POSSIBILITIES

CHAPTER 4

DESIDERATA FOR THE EVALUATION OF INFORMATION POLICY

There are several reasons for evaluating public policy. First, policies are formulated in order to achieve specific purposes. Evaluation of potential policy alternatives can support the decision process that eventually results in a concrete policy proposal. This type of evaluation can be undertaken by bureaucratic analysts, academic researchers, or stakeholders in the outcome of the policy.

Second, evaluation, that is used in the intermediate stages of the policy process between policy formulation and policy outcome supports the decision-making activity that legitimizes a policy in legislation or regulation; implements a policy under the constraints of time, resources, and ideology of the implementing agency; and monitors initial and intermediate policy outcomes with the purpose of fine-tuning a policy.

Third, evaluation assesses the success or failure of a policy once its final outcome is apparent. This type of evaluation can be directed to specific aspects of the policy, constrained by budgetary or other resources, limited to aspects of the policy that are quantifiable, assumed to be quantifiable (e.g., cost-benefit analysis), or focused on ideological and philosophical views implicit in the policy. Less often is a comprehensive review of the entire policy process undertaken to assess the cause for failure or success.

TYPES OF EVALUATION

Evaluation of a policy may be informal or formal. Journalists, broadcasters, and the citizenry at large undertake informal evaluation of some public policies. These are informal because they do not always enter directly into the decision maker's or stakeholder's ken. They are important to the entire policy-making enterprise, however, because they can indirectly affect decision making by rousing public opinion to support or to condemn a policy.

Formal evaluations, on the other hand, are undertaken either by direction of policy makers or policy implementers, or by academic researchers who specialize in the policy area under examination. These formal evaluators generally draw on many analytical techniques and approaches to carry out their formal evaluations. For example, they may carry out cost-efficient and cost-effectiveness studies, employ systems analysis, and gather information to answer questions that seem commonsensical relating to aspects of the problem that require a new program, probable success of the program, who will be affected by the program, whether the implemented program is consonant with the original ideas of the formulators of the program, and overall effectiveness (Rossi and Freeman, 1989, p. 18).

It may be assumed that the evaluator has some knowledge about the program to be evaluated and the general area in which the program intends to have an effect. A evaluator who is knowledgeable about toxic waste disposal, for example, would probably not be hired to evaluate a welfare policy, even though many of the techniques and information-gathering activities required by the two policy evaluations may be similar. Indeed, evaluative techniques, as valuable as they are, have their limitations. Many evaluative techniques are only considered successful within narrow parameters and only when issues of efficiency, effectiveness, and impact are germane (Rossi and Freeman,1989, p. 19).

WHAT IS THE INFORMATION POLICY PROBLEM, BROADLY CONCEIVED?

For some, the most desirable evaluation and proposed solution to an information policy problem would be akin to the diagnosis and repair of a faulty component of an automobile. If your steering wheel wobbles, the mechanic says: "Your tie rod is about to fall off. You need a new one. I will install it for you." You pay for the service, and the steering wheel ceases to wobble at least for a while. With such a problem, the mechanic's evaluation of the problem answered the questions of efficiency effectiveness and immediate impact. The mechanic's solution to the steering wheel wobble was efficient (a faulty part was replaced), effective (the steering wheel ceased to wobble) and had an immediate impact that was measurable (the car would steer reliably and a breakdown or loss of control of the car because of a faulty tie-rod would be avoided).

Information policy problems are not like that. What, then, are they like?

What is the specific problem that information policy attempts to solve?[1] To evaluate an information policy, a policy analyst must know what was intended by the policy and the rationale(s) that lie behind the specific policy actions carried out. Information policy establishes the parameters within which information is controlled (created, synthesized, analyzed, stored, disseminated, retrieved, and used) by human beings. If this is so, should not some discipline—such as informatics, perhaps—study the behavior of human beings and the means they employ to control information for specific ends? This discipline would be a source of knowledge for ways to control information, knowledge that could be applied to solve problems involving the control of information. Other human beings would then employ this knowledge in the solution of information control problems. This human attempt to solve information control problems is what is meant by information policy.

This neat and tidy statement begs many questions: What is the discipline that supplies this knowledge? Who will use the appropriate knowledge to solve these problems? What authority does this person or institution have to apply this knowledge? Is there other knowledge relevant to the solution of the problem, but which is not part of this specialized disciplinary knowledge? How reliable is either the knowledge from the discipline or the "other" knowledge of the humans who apply it? To what ends is the knowledge of information control being put? Does it promote justice or make a mockery of it? Does it help or hinder scientific communication? Does it incite sexual violence against women or act as a safety valve to inhibit such violence? Is the solution in accordance with current law? Is the solution economically efficient? Does the cost outweigh the benefit? These and other questions arise in evaluating or designing an information policy whose purpose it is to solve a problem of information control.

If information policy were carried out satisfactorily and accomplished its purpose, there presumably would be little interest in it. There is interest in information policy because: (1) information control problems have not been solved, and (2) there is a belief that these problems should be solved. These problems are often exceedingly complex. One goal for any information policy is that it should correct some defect that presently exists in that area over which information control is sought. This stance assumes that there is a solution to fix the problem. In fact, there may be no solution, but until policy makers are convinced of this fact, they will try to provide one.

There may be several reasons why information policy does not work. Specific policies may have overlooked certain contextual factors in the implementation of a policy; the initial premises in the formulation of the policy

---

[1]There is a danger in viewing a public policy as a technical response to a perceived problem. A "problem" is not amenable to a technical solution, but rather it is a dilemma that may only be resolved.

may be incorrect; the policy may not be implemented as designed; or there may be unforeseen problems or situations about which no one has thought. Almost all of these reasons for a failed policy, however, have a common denominator of knowledge. Do policies fail because of ignorance or because society is too complex and things do not turn out as they were intended? This question should make us pause to think about what expectations both policy makers and policy analysts have of information policies, for the way in which they evaluate them will probably conform to these expectations.

EXPECTATIONS OF INFORMATION POLICY

What do we expect information policy to do and how do we expect it to be done? The first response to these questions should be: "That depends on who 'we' are and what kind of information policy 'we' are talking about." The word "should" is used because if these questions do not arise, we—Americans living in the United States today —have no expectations. All information policies have a cultural, social, historical, and political context in which they are formulated. This recognition allows that if we were a different people, in a different time and place, our expectations of what information policy should do would differ. Further, by asking specifically what kind of policy we are talking about, the questioner also recognizes that what may be a legitimate approach to an information policy problem regarding pornography may not be appropriate for a scientific and technical information policy problem. Hence, calls for a national information policy that would cover the "information access" problem are either naive or misconceived. There cannot be a national information policy in the same way that there is a national energy policy. Information policy, broadly defined as a policy to control information, is simply too multifaceted and extends virtually to every aspect of life. Rather than having expectations of what information policy, in general, can do, other than to control information as intended, perhaps it would be wiser to ask questions appropriate to ask of any public policy. Stanley (1978, pp. 232–233) has posed these questions in his discussion of educational policy. When these questions are not asked, maintains Stanley, "policy talk will quickly wander into fanciful visions of social engineering." The questions are: Is it always appropriate to regard:

- Public policies as technical responses to "problems" that are open to technical "solutions"?
- The content of public policies as "social" policies addressing factors that affect only the social structural organization of society?
- Public policies as synonymous with government initiated and/or financed policies?

- Public policies as pragmatic "practical" responses to specific issues without bothering to debate them in light of evolutionary hypotheses regarding the likely directions being taken by the social and cultural order as a whole?

Stanley (Ibid.) comments on these questions by stating:

> Generally speaking, most policy discourse in which sociologists are being invited to participate is strongly biased toward affirmative answers to all these questions. It seems to me that as long as this bias continues, discourse about public policy will continue to generate the paradoxical reactions described earlier [on the one hand the increased talk of and demand for public policy and on the other hand the increased "public reaction against governmental intervention into the fabric of everyday life"]. The reason is that the public is being conditioned to demand from the government alleviation of its sufferings in technocratic forms that are illegitimate from the standpoint of the public's traditional understanding of freedom in a liberal, pluralistic, and democratic society like the United States.

He poses these questions in order to discover what it would be like not to have this bias. His questions with regard to information policies might be answered.

Is It Always Appropriate to Regard Public Policies as Technical Responses to "Problems" That are Open to Technical "Solutions"?

Technical policies, according to Stanley (1978, p. 233), are "activities regarded by their designers as technologically valid means for achieving a finite, specifiable end (the 'solution' of a 'problem')." Instead, he would opt for sensitizing policies, which are "activities intended to refine people's awareness that there are some problems that cannot (or should not be) specifiable as having solutions fully attainable by technologically operational means" (Ibid.). A policy that conceives of information as a manageable resource whose utility can be quantifiably assessed would be an example of such a technical policy. Office of Management and Budget (OMB) Circular A-130 (U.S. Office of Management and Budget, 1985, paragraphs 7c and 7d), for example, states:

> c. The free flow of information from the government to its citizens and vice versa is essential to a democratic society. It is also essential that the government *minimize* the Federal paperwork *burden* on the public, *minimize* the *cost* of its information activities, and *maximize* the *usefulness* of government information.
> d. In order to minimize the cost and maximize the usefulness of government information activities, the expected public and private *benefits* derived from

government information, *insofar as they are calculable,* should exceed the public and private costs of the information. (emphasis mine)

Those words that are considered the hallmarks of a technical policy have been italicized. Many will think that the primary sign of the technical policy is the phrase *insofar as they are calculable.* This is certainly the most blatant sign of technicism[2] in this policy. The other italicized words, however, in more subtle ways indicate the obfuscation that seemingly technical, neutral language brings. The triad of minimizing the burden, minimizing the cost, and maximizing the usefulness is what lies at the root of the debate of access versus dissemination with regard to this OMB policy. The words hide assumptions and values. What different stakeholders mean by these terms varies, in Stanley's (1978) terms, "with the focus of our moral attention" (p. 234). These terms obscure the value-laden problems with which a polity must contend when it is forced to decide what access to government information really means and what a polity wants it to mean. The terminology assures the "proper" functioning of a government department, not the "proper" access to government information produced by that department.

One may reasonably respond: What does one expect the department to do, give the information away? This is an appropriate, but not the most important question. No, one does not expect the department to give the information away. But the question that needs to be asked is: "What does 'minimize the federal paperwork burden' and 'maximize the usefulness of government information' mean?" These phrases conceal a whole set of assumptions and priorities. They are not just careless use of ambiguous language, masquerading as precise descriptions of definable processes and results. Uncovering these assumptions and priorities is one of the tasks of information policy evaluation. Once this is done, the question about giving the information away may be beside the point. The questions, then, turn to probing the legitimacy and the probable effects of actions implied, but never articulated by these ambiguous phrases.

Is It Always Appropriate to Regard the Content of Public Policies as "Social" Policies Addressing Factors That Affect Only the Social Structural Organization of Society?

Stanley (1978, p. 234) defines social policies as "activities intended to affect socially structured arrangements of some sort such as employment rates,

---

[2]Stanley (1978, p. 12) defines technicism as "a state of mind that rests on an act of conceptual misuse, reflected in myriad linguistic ways, of scientific and technological modes of reasoning. This misuse results in the illegitimate extension of scientific and technological reasoning to the point of imperial dominance over all other interpretations of human existance."

bureaucratic behavior, administrative organization, and financial patterns."
Cultural policies become "actions designed to affect the concepts and symbols that are embodied in social institutions and justify social practices." And,
further :

> Despite conceptual difficulties, the effort to distinguish between social policies
> and cultural policies is justifiable for many reasons. One reason is that the dis-
> tinction helps us to understand why many social policies are unsuccessful in
> fulfilling the goals that inspired them. What many people think are social prob-
> lems requiring social policies to solve them may in reality be cultural problems
> requiring normative changes in popular values and motives that cannot, in a
> liberal democracy, easily become a focus for official policies. (p. 235)

Are information policies social or cultural policies? Policies dealing with
obscenity, freedom of speech, and privacy are clearly cultural policies. They
directly affect, to one degree or another, the symbolic nature and moral ends
of information use itself. A good indication of this label of cultural policy is
the amount of controversy and passion that these policies engender in the
public at large. Citizens want some protection against unwanted intrusion
into their private affairs and they do not want anyone telling them what they
can or cannot say or what they can or cannot see. Public controversies over
the extent to which these claims are to be enforced by a public policy is a good
indication that information policies dealing with these areas are manifestly
cultural.

What about other types of information policies, such as telecommunications, scientific, and technical information policy? Are these social or cultural
policies?

One might first respond by doubting that these types of information
policies are social or cultural. Telecommunications policy, in general, focuses
on such things as the allocation of broadcasting channels within the frequency spectrum, the tariff structure that governs interstate telecommunications, or the services that various types of telecommunications service companies may provide to the public. Are these decidedly technical, social, or
cultural activities? Similarly for STI policy, is the control over the creation,
distribution, storage, retrieval, and use of scientific and technical information
a technical, social, or cultural activity? In these cases, the distinction seems
ambiguous. If citizens focus on the effect of these activities, however, they
have no choice but to admit that these are, at root, cultural policies. The
effects of these policies and the technologies that give rise to them, such as
long-distance communication capabilities or newly developed scientific instrumentation, are not neutral effects. They affect the way in which humans
perceive the world, and in which the world presents itself. One's personal
sense of control over the environment increases by virtue of the success of
these policies.

Information technology, and the policy that attempts to control its use, alter social relations and their moral ends. The invention of call waiting, to give a mundane example, affects the way a person assigns relative worth to telephone callers. The policy that allows this technology to be marketed to the public, and which is perceived by that same public to be progress, is the same policy that now allows a person to believe that one conversation partner can be dismissed because "a more important call has just come in." Prior to the introduction of this capability in our phone system, the only place one had to worry about such a decision was in face-to-face conversation. If another person interrupted me while I was talking to someone else, that person would be considered rude. If I dismissed my first conversation partner to talk to this second person, I would be considered rude. With this new capability, the question of my character or of the interrupter's character is not even considered. Call waiting is not necessarily bad; rather, both the technology and the policy that permits the use of call waiting (even if it is bowing to market demand) can alter cultural life. It affects, in Stanley's words "the concepts and symbols that are embodied in our social institutions and [which] justify social practices" (1978, p. 234).

In another way STI policy has come to be regarded as socially and culturally value-neutral. As indicated in Chapter 3, evaluations of previous STI policies revealed that the impact of a particular policy was rarely considered (Bishop and Fellows, 1989). Furthermore, those who recommended various actions to be taken to improve the flow of STI consistently focused on the organizational relationships and the organizational control of various programs. If only the *proper* agency, with *adequate* authority, could coordinate STI properly, the public was constantly told, then STI flow would improve and innovation would flourish. Policy analysts have failed to perceive the effect that this information and its efficient rational use has on cultural life. It can be used, on the one hand, to provide doctors with the means to ease suffering and cure disease and on the other hand to develop weapons of mass destruction that, in a pitifully naive brand of wishful thinking, are believed to be surgically accurate.

For those readers that are understandably squeamish with the notion of bureaucrats passing moral judgment and exercising control over the use of STI, one can only respond that they are already doing it through STI policies. The desire to prohibit access to information that is crucial to national security has been embodied in legislation, such as the Export Control Act and DOD directives that restrict attendance at scientific meetings where DOD-sponsored research is discussed (*Striking a Balance*, 1985; Gould, 1986). Those who protest against it argue for freedom of access, placing hope in the democratic process to control the social and cultural effects that it will have. Both sides by their very actions and positions on the relevant issues admit that these policies do affect one's view of the world and the ways in which cultural life may be changed because of it.

Is It Always Appropriate to Regard Public Policies as Synonymous with Government-Initiated and/or Financed Policies?

Stanley (1978, pp. 238–239) expands on this question, in this manner:

> The point of this question is not merely to ask where a policy factually originates, but rather where it should originate according to the political logic of the polity in question. When the people are sovereign in any seriously democratic sense, there are some policies that should not originate in governmental action (e.g. policies regarding the definition and enforcement of religious truth, censorship policies, etc.). Further some perhaps desirable policies could well be subversive of established state interests. A polity which forgets this is in danger of tyranny.
>
> The problem, of course, has become: how do we operationally specify the public will if not through the policies of democratically elected governmental representatives? . . . . There are some courses of conduct that are neither state policies, nor demands of "private" (parochial) interest groups, but are genuinely public commitments to certain common goods. These commitments are the stuff of traditions, of symbols, of the moral nuances present in the very uses of language.

For information policy in the United States, an awareness of the sovereignty of the polity is evident in the demand for both access and dissemination. The Freedom of Information Act is a policy whose intent, if not always its implementation, recognizes the polity's right to know, within specifiable limits, what the government is doing. Present-day controversies, however, seem more focused on the type and extent of services that the Federal government provides to the public. This is not to say that the government's response to the public demand for more information about its operations has been exemplary in each and every case. It has not. This is simply an assessment of the present-day information environment.

The problem with information services provided to the public, however, is that demands for private sector involvement in the provision of information services are not actively supported by any large public constituency. Their support is primarily from interest groups whose immediate livelihood is dependent on the absence of government from the information market. The issue over the privatization of the National Technical Information Service, for example, did not bring protesters to the streets or a ground swell grassroots movement called "SITiN for the NTIS!" Have we come to the point where either the polity is so heterogeneous in some respects that there are relatively few "public commitments to common goods," where government wishes to control the dissemination of public information, or where private interest groups are so powerful that public policy has become, in most people's minds, government-initiated and financed policies?

Is It Always Appropriate to Regard Public Policies as Pragmatic "Practical" Responses to Specific Issues Without Bothering to Debate Them in Light of Evolutionary Hypotheses Regarding the Likely Directions Being Taken by the Social and Cultural Order as a Whole?

Stanley (1978, p. 239) explains that:

> The pragmatist, impatient with theories and muddling through the political process, takes what he can get by way of achieved policies. To the theorist, the result is a patchwork quilt, promising much and delivering little, while exhausting the public's will to believe in rational progress.

Information policy, by and large, has been a pragmatic response to specific issues. The result, at least for scientific and technical information policy, is that "national scientific and technical information policy, like many broad areas of government concern, is not anywhere articulated in a comprehensive form" (Gould, 1986, p. 61). This lack of comprehensiveness is what Stanley (1978) would call the patchwork quilt perceived by the theorist. The alternative for the theorist is to broaden the framework within which policy debates occur and to suggest possible consequences to the cultural and social fabric that might ensue with specific policies. These possible consequences would encompass more than the ostensible goal of the policy (e.g., increased dissemination of STI), but would also include the potential social and cultural effects of the policy.

Some may argue that this has, in fact, been done with previous policies. Beyond the ostensible goals of STI policy lies the promise of economic development. To frame the social and cultural effects within the boundaries of economic progress, however necessary and desirable, would be to diminish the symbolic and moral nature of human existence. The pragmatic reader would probably respond by saying "Poppycock!" or whatever it is pragmatists say when they believe that the theorist has gone too far. Perhaps it is poppycock. The alternative, however, is not to raise the issue at all and be convinced that policy makers' propaganda about the nation's desired future is, in fact, true. On the other hand, the theorist may also be incorrect. There must be a tension maintained between the practical demands of the pragmatist and the ideals of the theorist. If the polity lets the tensions subside by acquiescing to either, human dignity and autonomy will suffer and information policy can turn into an instrument of tyranny.

Once policy analysts and policy makers have asked these questions about information policy, should not their expectations broaden to include other areas and concerns that they had previously ignored? Undoubtedly, they should. Once sensitized to these other areas and concerns, they are then prepared to ask the fundamental question of evaluation: Does the policy do what it intends to do? There are many ways to answer this question. One way that has predominated in STI policy was to say that the policy was not effective because an agency possessing legitimate and adequate authority did not carry out the policy. If policy implementation were to be moved to agency X or if agency Y were to be created, then the policy would work.

This type of evaluative analysis focuses solely on the organizational aspect of the policy. It neglects not only the cultural effects that the policy might have had, but also specialized knowledge about information control. To correct this tendency, the case study presented in Chapters 6 to 8 attempts to ascertain the reason or rationale that lies at the base of specific recommended policy actions. "Experienced professionals" in information science examined these reasons to see if the discipline of information science, as they knew it, would support or undermine the reason given or whether it was relevant at all. Doing this for several recommendations maps some of the areas to which information science contributes its specialized knowledge of information control. Furthermore, the failure of a policy cannot always be attributed to organizational shortcomings.

The case study is not exhaustive, but it serves as an example to show how policy analysts might determine which specialized knowledge is applicable for information policy dilemmas. Even if they can identify specific applications of information science in information policy, does this mean that they possess a knowledge base by which even the pragmatist can muddle through and enact policies that are effective? What kind of knowledge would they need? This question leads to a discussion of the role of prediction and the type of knowledge needed in information policy.

PREDICTION AND ANALYTICAL ANALOGS

In information policy, encompassing as it does both technology and society, policy analysts risk downplaying the problem of the applicability of specific knowledge to policy problems. With certain technological aspects of information policy, such as specifying standards for coaxial cable, engineering disciplines can predict the reliability of the cable within specific tolerance limits, under certain conditions. On the other hand, the allocation of the frequency spectrum in telecommunications policy may seem fairly straight-forward, but other factors—political, ethical, and sociological—will and should affect the allocation decision. These factors will erode the predictive nature of any disciplinary knowledge. If policy makers attempt to allocate the spectrum only according to some predetermined formula that allots the number of channels based on current usage they will be applying predictive knowledge, but it may not be applicable to the entire problem as the example with the Kingdom of Tonga showed in Chapter 2.

In any social science analysts may not be able to apply knowledge that is predictive in a closed microsystem to a broader open macrosystem. In discussing the predictive ability of social science, Rein (1976) has said:

> While one of the crucial tests of the objectivity of understanding about social events is the confirmation of predictions based upon it, this is seldom possible in the social sciences. Still, when someone consistently gives good advice, we trust their understanding of events and their capacity to translate their understanding into practical advice giving. Thus, if a person's predictions are generally realized, we believe that their understanding is objectively grounded. (pp. 261–262)

Further, Rein (1976) states:

> I believe that efforts to imitate the analytic techniques of the natural sciences as a way of providing reliable predictions are based on a misleading analogy. First, pure science depends upon analytic procedures which are not possible in the study of society; it deals with presumably stable and universal relationships between events, while social events are neither stable nor universal, and it can discriminate disaggregated factors and re-combine them in a way which seems beyond the reach of social analysis. But secondly, we are not, I think, ever really interested in 'pure' social science—a social science divorced from action, since the meaning of social events is inextricably bound up with the values we attach to them. I think that we accept the apparent relevance of pure science because the categories which define the natural world seem meaningful, irrespective of values. We assume that any understanding of the natural world is potentially useful, even if we cannot see how it might be used. (p. 264)

Rein (1976, p. 264) also rejects an analogy based on applied science because "it is hard to predict exactly how that machine will function in an environment. . . . Even if the machine works, it may not be as valuable socially as expected—for instance, it may have unforeseen and undesirable social consequences."

The difficulty in information policy is determining which analogy is appropriate not only within specific policies, but within specific policy tasks. For many suggested policy actions, neither the natural science nor the applied science paradigm may be applicable. In such cases, policy analysts must be concerned with what Rein calls "the overall complex of behavior." He explains that:

> This understanding cannot be reached by aggregating known regularities in the behavior of discrete events. Rather, social understanding in this situation depends upon telling relevant stories: that is, deriving from past experience a narrative which interprets the events as they unfolded and draws a moral for future actions, suggesting, for example, how the future might unfold if certain steps were taken.
>
> The giving of advice and the design of social programmes is like the telling of relevant stories. Such stories resemble proverbs and metaphors, for they seek to match reality to archetypical patterns of events by drawing analogies. That is to say, they provide an interpretation of a complex pattern of events with normative implications for action, and not with a universal law. Nor is the correspondence expected to be much more than a warning and a direction. (Ibid., pp. 265–266)

So beyond identifying knowledge applicable in an evaluation of an information policy, how should that knowledge be applied? Not only must policy analysts develop a way to analyze information policies to decide what disciplinary knowledge is applicable, but having once determined that, they must further decide if the relevance of this knowledge is to be explained by an applied science analogy or by Rein's social science analogy.

DESIDERATA FOR THE EVALUATION OF INFORMATION POLICY

After all this preliminary posturing this chapter describes what analysts would like the evaluation of information policy to accomplish. These desiderata can be discussed under the rubrics of comprehensive description, multifacetedness (nonreductionism, politics, economics, law, efficiency, impact, ethical concerns, and cultural effects), policy actors and stakeholders, policy goals, resources, policy results, legitimacy of rationale for actions prescribed (based on empirical knowledge and theoretical knowledge), and suggestions for future action.

Description

An evaluation should describe the general outline of a policy and the background that gave rise to that policy. This includes such matters as previous policies that have been promulgated in this general area; the situation that gave rise to recognition of the problem addressed by the policy; the cultural, social, and political context in which the policy is carried out; and the results of similar policies.

Multifacetedness

This part, an extension of the description of the policy, becomes more specific by addressing several aspects of the policy and its context in more detail. For example, care should be taken to avoid reducing the policy problem to that of economics, politics, organizational management, efficiency, or cultural aspects. Most policy problems include all of these aspects. By reducing the problem to one of these aspects, policy makers will be prone to find some type of technological fix to the problem, a fix that will have other unintended effects.

Policy Actors and Stakeholders

The evaluator must identify the various policy actors and stakeholders of the policy. Policy actors are those who formulate or implement policies; stakeholders are those whom the policy affects. Predictions about human behavior are often erroneous. Policy analysts should not predict that actors and stakeholders will act a certain way simply because they wish them to do so. In postimplementation evaluation the evaluator can examine the actions of the various actors and stakeholders in order to determine how their behavior influenced the success or failure of the policy.

Policy Goals

Policy goals must be identified in order to determine if the actions proposed or actions taken met these goals. Are the goals clearly and unambiguously stated? Do the goals hide a series of assumptions and value judgments that the language in which they are stated does not necessarily imply? (Note the example of OMB Circular A-130, discussed in this chapter.)

Resources

Did the policy actors have adequate resources to carry out the goals of the policy? Did policy actors need authority that was not given them? If expenditures and personnel were involved, did the policy accurately foresee the amount of money and number of people needed to accomplish the goals of the policy?

Policy Results

Did the policy achieve what was intended by the goals? What other effects, such as cultural, social, economic, and political, did the policy have? Who was affected by the policy?

Legitimacy of Rationale for Actions Prescribed

Was the rationale for actions taken in the policy supportable by common sense, empirical knowledge, or by theoretical knowledge from a relevant discipline? What disciplines or bodies of knowledge provided the rationale for suggested policy tasks? Could the policy problem be solved in the first place or was a solution forced on an insolvable, but only resolvable, situation?

Evaluator Self-Critique

An all-encompassing evaluation may not be possible given constraints of time, resources, or institutional organization and goals (Feldman, 1989). Should not the evaluator make clear what was sacrificed and what views were relegated to subordinate status? In short, should not the evaluator reveal his or her own biases?

Suggestions for Future Action

What steps can be taken to correct or mollify the failures of the current policy? What rationales are used to make such suggestions? What disciplinary knowledge supports these rationales?

SUMMARY

This chapter describes the elements involved in evaluation by defining what evaluation research is and the types of evaluation that can be employed. It examines the information policy issue (the control of information) and expectations of that policy (Stanley's four questions). The chapter also provides a discourse on prediction and analytical analogs, and a short list of elements every evaluation should include. The chapter undoubtedly poses more questions than policy analysts have answers at the present time. The questions add new dimensions to analysts' and policy makers' perceptions of information policy problems. Policy makers do not just have economic or legal principles at stake when they conceive of information policy solutions. The solutions affect cultural life in ways that they can only dimly perceive prior to the policy and only slightly more clearly after implementation.

Using Aristotle's inductive axiom that the person who understands theory is the best practitioner (Taylor, 1955, p. 29), the next chapter examines currently held concepts of information science and information policy in order to understand their aims. Regarding information science, the chapter will focus on the types of knowledge peculiar to this discipline, the boundaries of the discipline, the assumptions basic to it, and speculations on its applicability to information policy. For information policy in general this chapter will examine widely held assumptions of what it is, what it intends to accomplish, and what problems are attendant in it. Subsequent chapters offer a case study supporting the thesis that information science at present has limited applicability to macroinformatic problems and is not equipped to answer many of the questions posed and caveats given in this chapter.

CHAPTER 5

INFORMATION SCIENCE AND INFORMATION POLICY: A CURRENT ASSESSMENT

Machlup and Mansfield (1983, pp. 4–5) in their wide-ranging investigation of the study of information have wrestled with what information is. They conclude that:

> Information is not just one thing. It means different things to those who expound its characteristics, properties, elements, techniques, functions, dimensions, and connections. Evidently, there should be something that all the things called information have in common, but it surely is not easy to find out whether it is much more than the name.

Previous chapters have defined information policy as policy that attempts to control information. A healthy-minded skeptic may ask that if information can be defined in so many ways, should not one be more precise about what it is one hopes to control? The skeptic certainly has a point and the avenue suggested by the skeptic is certainly one way to approach the problem of information policy. A rejoinder that now seems convincing is that in spite of the academy's inability to define information precisely, policy makers have not been hesitant to formulate information policy. Either defining what information is has little to do with making information policy or one of the reasons why there is a problem with information policy is that academic theorists and policy analysts have not been able to define with sufficient precision those phenomena over which policy makers seek control with their policies. How can this impasse be resolved? At this point there does not appear to be a clean, elegant way of doing it. Instead, policy analysts must make what may seem like bold, and often foolish, attempts to deal with the impasse in a practical way, by making suggestions that may prove to be either persuasive or not persuasive. These suggestions may then be refined in the course of analysis and trial to formulate other suggestions and possibly other rejections. The other alternative does not seem viable. If those involved in the information policy making enterprise simply remain where they are and conjure up more hairsplitting definitions then progress in understanding either the phenomenon of information science or information policy will not be made.

This book attempts to lay some groundwork for just such a practical foray into making some suggestions about the disciplinary knowledge of information science and its relation to information policy. This chapter will first examine some contemporary ways that information science is perceived by its practitioners. This section will directly and indirectly answer such questions as: How is information science currently defined? What are the boundaries of the discipline? What are its aims? What type of knowledge does it encompass? What basic assumptions are made about its applicability to practical, concrete problems? This chapter provides a similar examination of information policy: What are the aims of information policy, what are the areas over which it seeks control, what is now lacking in information policy making, and what are the assumptions made about the knowledge necessary to make such policy?

INFORMATION SCIENCE

Definition

In an effort to define the discipline of information science with some rigor, several different approaches have been taken. Probably the most voluminous of these is the terminological approach. Schrader (1984), in a summary article based on his dissertation (Schrader, 1983), "attempts to set forth the chronology of disciplinary names which have been used in the literature over the past 80 years to characterize information science and its conceptual antecedents." His summary of past definition attempts is germane:

> No consensus has emerged in the scholarly community with respect to the scope and nature of the domain. Terminological chaos is the description most often used to characterize the plethora of definitions of information science which have been set forth in the literature over the past 20 years. The chaos in terminology reflects the chaos in conceptualization which has dominated not only information science but the earlier efforts to differentiate the domain from traditional librarianship. (1984, p. 237)

This characterization is the most accurate description of the existing literature. Mikhailov, Chernyi, and Giliarevskii (1984, pp. 363–383), in a similar manner but with much less rigor, explore the history and development of the various terms that have been used in denoting this field in the United States, Germany, France, and the Soviet Union. Houser (1988), in analyzing articles submitted to the *Journal of the American Society for Information Science* concluded that:

> *JASIS* papers provide no evidence of disciplinary aspirations. Hence, the claims of inter- and multidisciplinarity which figure prominently in the definitional literature are merely assertions without any intellectual basis. If logic prevails at all, then the neologism "information scientist" is also without any validity. (p. 28)

Other authors, however, have taken less direct approaches in order to come to grips with the disciplinary boundaries. Salton (1985), in response to a letter by Keren (1984), deals with the relationship between information science research and practice. Keren charged that the discipline of information science was not progressing. He said:

> Although research in information science is not, perhaps, that stagnant it would be worthwhile to find out how much of it has really contributed to our body of knowledge and to the methods used by practitioners. I dare say that we will probably be rather disappointed. I did not research this, nor do I have firm data, but on the strength of my own experience, as a practitioner, I have the impression that only very rarely useful and nontrivial information emanates from the research published in our professional journals, including those who strive to set high standards. (Ibid., p. 137)

Salton (1985), in turn, presents nine categories of active research areas in information science, thereby invalidating Keren's claim about the moribund nature of the field. He does not attempt to take the next logical step to use these areas to sketch the perimeters of the discipline.

Boyce and Kraft (1985), however, do take this next logical step by trying to identify principles and theories of information science. In their formulation, a principle is "a fundamental law; generally an empirical regularity based on continued observation" (p. 154). A theory is considered

> to incorporate a body of such principles and to suggest new principles that can be tested as hypotheses, both to increase knowledge and to invalidate or to strengthen the basic theory itself. Thus, a theory is not a synonym for impracticality, as some would seem to imply. (p. 154)

Boyce and Kraft (Ibid.) admit that although there may be many principles in information science, no theories exist, as defined above. Although "they offer useful explanations of empirical principles, they do not necessarily lead to new knowledge. . . . We see our discipline as primarily practical and technological" (p. 155).

Their review identified four areas that provide useful principles. These are information theory, representation for retrieval, models of information retrieval, and bibliometrics, outlined in detail as follows:

(1) *Information Theory*: Information theory is based on the model of Shannon and Weaver (1949). The general study of communication, especially scientific communication, is an outgrowth of this model. This general communications model, which includes a sender, a message to be sent, a channel along which it is to be sent, a receiver of the message, and the effect of the message, has been used widely in areas such as studies of mass communication and biological studies of human information processing.

(2) *Representation for Retrieval*: This area covers the ways in which items are to be retrieved and represented within an information retrieval system. Its main focus is the design and evaluation of indexing languages and the extent to which index terms are representations of classes of actual documents.

(3) *Models of Information Retrieval*: These models are actually "attempts to relate representations of questions or queries for information to representations of records in a database, usually in terms of bibliographic references to documents (Boyce and Kraft, 1985, pp. 160–161). The relevance of a document to a query and the use of logical, probabalistic, or other models in this judgment are examples of the areas covered.

(4) *Bibliometrics*: This is "the quantifiable study of written communication through its physical realizations." (Ibid., p. 163) Several bibliometrical laws, such as Zipf, Bradford, and Lotka, are well known, but Wallace (1985, p. 36) contends that "although there have been some tentative attempts made to establish a theoretical base for bibliometrical studies, no clear explanation for any bibliometrical phenomenon has been developed."

Principles and techniques falling outside these areas, even if they are related (e.g. information economics), are often not considered information science. However, the listing of the above areas does not exclude the study of scientific communication, for example, which uses principles and findings from all four areas.

Case (1985), in a similar vein, discusses four broad areas of information science activity: information retrieval and indexing, the principles of economics of information, the automation and use of information services, and information technology. His primary conclusion is that information science "is a maturing discipline" (p. 49). His method of dealing with the discipline, while interesting, is little more than a congratulatory song of praise to the profession.

All these definitions and descriptions given so far do map out a distinct area of investigation. Most writers have tried to define clearer boundaries of the discipline within this large area. For the purposes of the case study presented in subsequent chapters, however, what was needed was not a narrow working definition of the field, but its opposite. Such a definition should do little more than sketch boundaries. In so doing any information scientist's mental model of the field would most likely be more narrow than the definition used here. Here information science is broadly defined as that discipline dealing with problems associated with controlling (collecting, preparing, analyzing, translating, synthesizing, storing, disseminating, accessing, or retrieving) information or surrogates of such information, such as abstracts; indexes; bibliographic citations or reviews; or the collection, preparation, storage, dissemination, and retrieval of information about such information through bibliographies of bibliographies, guides to the literature on a specific subject, etc.. Described in this way, information science is similar to the engineering disciplines.

Another way of describing the field is to look at the institutions and people involved in its history, as was carried out by Herner (1984). After looking at seminal writings, major actors, and major events and developments, he summarizes his study by saying:

> Information science is the product of convergences of library science, computer and punched card science, R&D documentation, abstracting and indexing, communication science, behavioral science, micro- and macro-publishing, video and optical science, and various other fields and disciplines. (p. 157)

Exactly what this term "convergence" actually means (see also Rayward, 1983a) is not clear. Herners's summary statement echoes an article by Foskett (1970), who affirms Herner's perception (although Foskett wrote 14 years earlier). According to Foskett (1970, p. 365), "A *new* discipline emerges, tentatively and shakily, from a synthesis of parts from diverse fields into a new, coherent whole, and not as a mere enlargement or improvement of parts, or even of an existing whole."

His Saganesque tone notwithstanding, Foskett's characterization of the discipline still seems to hold. It is significant, and at the same time troubling, that after so many years little progress has been made in discerning disciplinary boundaries. The discussions of the differences between information science and library science (Rayward, 1983a) essentially arrive at the same conclusion: The field is not well defined and at this point probably cannot be well defined.

Aims of Information Science

Machlup and Mansfield (1983, p. 18), in attempting to characterize the various disciplines covered in their book, report that information science has several aims, dependent on the area of applicability. These aims are outlined as follows:

(1) In its broadest sense, it stands for the systematic study of information and may include all or any combination of the academic disciplines discussed in this volume.[1]

(2) When included in the phrase *computer and information science*, information science denotes the study of the phenomena of interest to those who deal with computers as processors of information.

(3) In *library and information science*, it indicates a concern with the application of new tasks and new technology to the traditional practices of librarianship.

(4) In its narrow sense, information science is used as the name for a new area of study that is evolving from the intersection of the other three mentioned areas, with perhaps a special interest in improved communication of scientific and technological information and in the application of well-tested research methods to the study of information systems and services.

With these descriptions, the aims of the disciplines described in general terms here could be seen eventually to cover the control of all human knowledge. In fact, Flynn (1987, Chapter 1) approaches information science as if it were the discipline to be applied to question answering of all kinds. His paradigm of collecting the data, choosing a technology, choosing the relevant data, coding the data, manipulating the data, using the data, problem solving and decision making, establishing the values, weighting the criteria, data display, implementation of the chosen alternative, and evaluating the information is one that could be applied to any question in any discipline. Flynn implicitly raises the question of whether information science is an all-encompassing science. Mikhailov et al. (1984, p. 382) call this a "patent absurdity," because

> The task of informatics[2] is not the creation of methods for the logical processing of scientific information from the various fields of knowledge in order to obtain new information that was not directly given in the original information. It is apparent that such logical processing of scientific information and its qualitative

---

[1]Cognitive Science, Informatics (Computer and Information Science), Artificial Intelligence, Linguistics, Library and Information Sciences, Cybernetics, Information Theory, and System Theory.

[2]Informatics has a more restricted definition than information science. For Mikhailov et al. (1984, p. 365), informatics "is a scientific discipline and not an independent branch of science . . . studies the structure and general properties of *scientific* information, not *all* information and not *semantic* information . . . is concerned with the study of all processes of scientific communication that goes through formal channels (i.e., the scientific literature) and information channels (i.e., personal contacts between scientists and specialists, correspondence, exchange of preprints, and so on.); . . . belongs to the social disciplines, since it is concerned with the study of phenomena and laws peculiar to human society." In summary, informatics is "that scientific discipline that studies the structure and general properties of scientific information and the laws of all processes of scientific communication."

evaluation are impossible without the use of the facts, laws, and theories of that science to which this scientific information belongs.

Mikhailov et al. (1984, p. 378) question whether informatics, as a body of knowledge, has even been formed as an "independent branch of science." They state:

> For establishing such signs, it is necessary to define the subject of research, work out ideas that are appropriate to this subject, establish a fundamental law that is inherent in a given subject, and discover a principle or create a theory that permits explanation of the multitude of facts.
>
> For informatics, only the first two conditions have so far been met. This means that informatics has not yet emerged from the formative stage and is approximately in the same situation as cybernetics, semiotics, structural linguistics, and other new scientific disciplines. Since informatics has not completed its formation, its subject is constantly being refined.

In spite of this constant refinement, it would be worthwhile to look at some recent examples of information science in action.

What Types of Knowledge Does Information Science Embrace?

Throughout this examination of information science, the practical aspect has constantly been stressed. In fact, if several recent books on information science are examined, this practical aspect will stand out all the more. This examination will also help to answer the question: What types of knowledge does information science embrace? The answer will be tied to the three types discussed in Chapter 4: scientific knowledge, engineering or technical knowledge, social science knowledge, or humanistic knowledge.

In their history of 40 years of information science, from 1945 to 1985, Lilley and Trice (1989) have approached the topic from the point of view of the most active areas in information science and the personalities active in those areas. What results is a characterization of information science that is extremely practical. After an initial period in the development of nonconventional information systems (1948–1968), information science has been applied to libraries, to online databases, and networks among libraries and any other interested organization. This type of knowledge certainly seems to be of the technical or engineering variety, although it has encompassed the other types from time to time. By and large, however, it is a discipline that has enabled information specialists to get things done through knowledge representation and retrieval of information. This has been done under the systems umbrella by using a bibliographic system, usually within one institution, as the area for study. The network activity has sought to bring all these individual retrieval systems together for everyone's mutual benefit.

In short, the problems attacked by information science have been micro-informatics problems (Lancaster and Burger, 1990) and not macroinformatics problems. Networking, the lone exception to this, has not moved (and perhaps may not move) beyond both the concept of the possibilities inherent in the connecting together of individual systems within single organizations and the technical and administrative problems related to this endeavor.

Olsgaard (1989), in a recent collection of papers, expands beyond the problems of retrieval and networks to encompass not only the arrangement and retrieval of information, but also information science theory (information theory, bibliometrics, citation analysis, and linguistics) and the practice of information science in library organizations (organizations and information systems, administration of the library automation process, and the measurement and evaluation of information services). This is helpful, but still relegates information science applications to the realm of microinformatics.

Finally, Losee (1990) charges that:

> Some scholars calling themselves "information scientists" are justifiably more concerned with specific information-related problems or environments than with the methods used to study information. They have let these practical problems lead them away from scientific and appropriate methodologies into providing incorrect solutions for problems because of the pressure to find a solution. Information science has not moved very far in the hands of people with this attitude. Many of these people would more appropriately be called "knowledge technologists." They focus on knowledge, a structuring of information, rather than on the information itself, and they focus on the technology associated with storing and manipulating knowledge rather than the scientific examination of information. (pp. vii–viii)

Losee, in contrast to these scholars, deals with information and its measurement. He confidently believes that "information can be measured and that information science is a viable discipline which can describe and predict the actions that take place in what we commonly consider information production, use, exchange, and retrieval" (p. ix).

This would seem to imply that the science of information (equals information science?) is a scientific discipline, in the sense used in Chapter 4. Most theoreticians and practitioners, however, in spite of the fact that information science involves the control of information by humans, would not go so far as to say that information science is a social or humanistic discipline.

While all this speculation about the nature of the discipline is both interesting and necessary, it often does not move beyond the seeming confusion described here. Kochen (1983), in spite of his extreme views (see the critique by Rayward, 1983b) does help to sum up the point of view presented in this book. He says, "What matters most is that investigators who identify with the information disciplines formulate researchable problems and make discoveries, and contribute insights that clarify the nature and dynamics of information and knowledge" (p. 371).

This practical and clear-headed statement is warrant enough to explore the relation of information science to information policy.

INFORMATION POLICY

Information policy is an area whose importance for society is often emphasized, but one in which little analytical work has been accomplished. In a recent paper, Heim (1986, p. 21) thought it imperative that information scientists in the United States "understand and influence the complex factors that comprise the contemporary information infrastructure." In a similar vein, Sherrod and McFarland (1976, p. 73) cite the need for "a real understanding of the nature and needs of the information user." These two complementary views, stated almost 10 years apart, suggest that the basic knowledge underlying intelligent information policy making is lacking. This can be applied both to U.S. and to non-U.S. information policies. The type of knowledge called for seems to be macroinformatic, social science, or humanistic knowledge, not the scientific and engineering type of knowledge by which the discipline of information science is often described. Could it be that there may be little relation between information policy on at least a national level and the discipline of information science? Another possibility is that policy makers have been formulating information policy on a national level using knowledge available from information science that is apparently valid on a local, institutional or microinformatic level. There seems to be an area of ignorance here that has not yet been explored.

There have been several attempts, however, to discuss and analyze information policy in spite of this apparent ignorance. One approach is the classification of information policies into categories based on the goal of the policy. Chartrand (1986a) and Milevski (1986), for example, have used the same categories to classify legislation dealing with information into nine categories:

(1) Federal information resources management;
(2) Information technology for education, innovation, and competitiveness;
(3) Telecommunications, broadcasting, and satellite transmissions;
(4) International communications and information policy;
(5) Information disclosure, confidentiality, and the right of privacy;

(6) Computer regulation and crime;
(7) Intellectual property;
(8) Library and archives policies; and
(9) Government information systems, clearinghouses, and dissemination.

Although this approach is extremely useful in helping policy makers and policy analysts to perceive the wide range of policies gathered under the rubric of information policy, it only deals with these policies in a superficial way. Such a classification does not help them to understand, for example, the subtle differences in the goals of these policies, the effect of a policy's content on its formulation and implementation, or, beyond the method demonstrated, any other way of analyzing or evaluating information policies.

Another approach is that demonstrated by Burger (1986), who discusses the political factors surrounding the formulation and implementation of public policies in general. He suggests that the factors he identifies be studied in regard to existing information policies. This approach also falls short because whereas it focuses on an important area of information policy analysis, it only deals with the politically relevant influences on information policy and not with the broad complex of behavior that underlies information policy. Linowes and Bennett (1986) also deal with privacy policy in this particular, restricted way.

Still another approach that seems to be attracting more adherents is a reductionist one. (See, for example Bushkin and Yurow, 1981, Braman, 1988.) In this type, information policy is reduced to exclusively economic or legal terms and judged accordingly. While this does overcome the vagueness of the two other approaches, it restricts analysis and evaluation to primarily economic or legal terms and ignores, for the most part, noneconomic and nonlegislative aspects.

A fourth category of information policy analysis can be called the issue identification and option approach (Jacob and Rings, 1986; Bikson, Quint, and Johnson, 1984; McIntosh, 1990; Yurow et al., 1981). Such writings deem it sufficient to do little more than enumerate the issues of information policy and call for their resolution. This is done, unfortunately, with little analysis of the issues so identified. This approach, like Chartrand and Milevski's, mentioned above, is certainly useful, in helping policy analysts to realize the dimensions of the problem; It does not, however, provide them with a coherent framework for evaluating such policies.

Fifth, there is the large system approach best exemplified by *National Information Policy* (1976). This document attempts to identify primary sectors of information policy in the United States and suggests a coordinating agency to oversee the resolution of information policy problems. In spite of its designation here, however, systems analysis is not used to understand the interaction of the sectors identified. Although this document has now achieved the status of a classic, and while some of its insights are valuable, it does not seem to have moved policy analysts closer to understanding or resolving information policy problems.

Sixth, there is the private stakeholder approach (Bezold and Olson, 1986) whose goal it is to "focus in depth on the long term direction of our industry and the public policy framework in which it can most effectively operate for the good of all" (p. 1). Bezold and Olson (Ibid.) offer four "scenarios" that the future may take. They describe three basic characteristics of information policy issues that appear to take a broad, overarching view of the entire problem, similar to the view advocated in this book. They state:

> Information policy issues have three characteristics that, while not unique to this policy area, are important to keep in view. First, information policy is not guided by a national overarching goal, like "energy independence," or by a coherent national plan, like the plan for interstate highways.
>
> Second, information policy issues contain large, inherent uncertainties about technology and market behavior. Large uncertainties can lead reasonable people to quite different judgments about the nature and seriousness of issues and the most effective ways to resolve them.
>
> Third, information policy choices are usually not between "good" and "evil," but between legitimate and competing values, goals and interests. As a result, these issues are not likely to be resolved completely in favor of any polar position. The challenge at any particular time is to strike an "appropriate balance" among the conflicting values and interests.
>
> These basic characteristics have an important implication: the long-term resolution of information policy issues must involve the participation of all stakeholders' points of view. (pp. 5–1, 5–2)

Other studies have focused on certain aspects of information policy at the national level. For example, Porat and Rubin (1977) have provided a macro-informatic description of the information economy. *Intellectual Property Rights in an Age of Electronics and Information* (U.S. Congress. Office of Technology Assessment, 1986) has explored the problem of intellectual property rights and the problems of assuring these rights in the present computerized environment, and *Informing the Nation* (U.S. Congress. Office of Technology Assessment, 1988) has looked at the dissemination of Federal information. The National Commission on Libraries and Information Science (1982) looked at the interaction of the public and private sectors in providing information services. While these approaches have been helpful, they still have failed to provide a framework within which to analyze an individual information policy and to provide a more complete understanding of this phenomenon.

Only recently has there been any progress made towards providing ways of analyzing information policy and providing frameworks for this analysis. McClure, Hernon, and Relyea (1989) not only give the historical development of U.S. government information policies, provide several perspectives from which to view them (Office of Management and Budget, congressional, private sector, and citizen), and outline key policy areas, but they also demonstrate the value of various types of models to provide frameworks by which to analyze policies. These authors (see also McClure, 1989; Hernon and McClure, 1987) have focused specifically on policies governing the control of government information.

Most other writers have not been as precise in their definition of information policy nor the domain covered by it. Rosenberg (1981, p. 3), for example, says that the purpose of information policy is "to govern the creation, distribution, and use of information." Chartrand (1986b, pp. 6, 9) defines information policy as "policies which govern the way information affects our society." Hayes (1985, p. 17) argues that information policy "is the basis for societal and institutional decisions concerning the allocation of resources to the acquisition, processing, distribution and use of information." Mason (1983, p. 93) suggests that:

> In the United States, 'Information policy' is actually a set of interrelated laws and policies concerned with the creation, production, collection, management, distribution and retrieval of information. Their significance lies in the fact that they profoundly affect the manner in which an individual in a society, indeed a society itself, makes political, economic and social choices.

What this disarray of approaches, circular definitions, and quasi-analyses indicates, it seems, is that information policy, whatever it is, is exceedingly complex. Present analyses have not yet brought analysts and policy makers to the point of understanding, much less influencing, the phenomena over which information policy is designed to exert control. Successes have been, generally speaking, in the area of political analyses (Linowes and Bennett, 1986), macroeconomic description (Porat and Rubin, 1977), or other partial approaches. Present knowledge, however, does not allow policy analysts to evaluate a policy proposal or existing policy with any reliable, verifiable method. One concept, although nowhere fully articulated, does emerge, however. Information policy requires several different disciplines for evaluation: economics, law, political science, public administration, sociology, public policy, management science, and information science. These disciplines have all been used or alluded to in many of the papers cited. (See, for example, Bushkin and Yurow, 1981 for law and economics; Linowes and Bennett, 1986, for political science, public administration, and public policy; Hayes, 1985, for management science; and Rosenberg, 1981, for information science.) One challenge is to learn how to evaluate policies using these disciplines without allowing one discipline undue predominance over the other. Chapter 4 considered just this problem.

Another danger is that policy makers and policy analysts know so little about individual information policies that their present attempt to analyze or describe a comprehensive information policy is presumptuous. Information policy can legitimately be used only as an umbrella term for a group of public policies united in one way or another by that ambiguous term "information." This underscores the unsatisfactory definition of information, reluctantly recognized in the first section of this chapter.

SUMMARY

The previous analysis has attempted to give a broad-based assessment of information science as a discipline and of information policy as a coherent policy area. A pessimistic reader may wonder if such a complex and seemingly hopeless task was necessary. Both information science and information policy are areas in need of development and new approaches. According to Kochen (1983, p. 371), researchers need to "formulate researchable problems and make discoveries, and contribute insights that clarify the nature and dynamics of information and knowledge." In an effort to take his advice, the next three chapters present a case study of a scientific and technical information policy proposal, the SATCOM report, that was issued in 1969. The case study probes the relationship of information science to this particular information policy by asking practitioners in the field whether principles of information science lie at the base of policy recommendations.

PART III

CASE STUDY OF SATCOM REPORT

CHAPTER 6

APPLICATION OF
INFORMATION SCIENCE
TO SATCOM: PROBLEM
AND METHOD

THE PROBLEM

Some attempts to provide a broader context for studying information policies have been made. These, however, have not addressed specific questions relating to the knowledge necessary to evaluate information policies, nor have they provided any insights into the problem of understanding the complexity and multifacetedness of information policy in general.

In assessing the state of research on information policy, Trauth (1986, p. 42) has stated that "as information policy development has been technology-driven, policy research has been discipline-bounded." By *discipline-bounded* she means that researchers have only examined those information policy problems that are immediately relevant to their own disciplines. As a result of this narrowness of attention, she maintains, information policy research has remained fragmented; those whose research interests focus on one area of information policy rarely adopt research strategies of researchers in other areas, and findings in one area often are not applicable in another. As an example, Trauth cites the research activity of those within computer science and business administration—activity that has increased because of auto-mated recordkeeping practices. Cryptology and the work of the Privacy Protection Commission are obvious examples. Computer science researchers have developed sophisticated algorithms to encrypt data in telecommunications networks. This type of research is primarily technological and scientific in nature. The work of the Privacy Protection Commission, on the other hand, focused on constitutional and commercial law issues raised by the introduction of new technology. The Commission's work was primarily political and legal in nature. Although research in encryption and in political science and law may be related, investigators with different disciplinary training do not usually contribute to each other's discoveries in these areas.

Trauth develops what she calls an integrative approach to information policy analysis. This approach involves rejecting the "technology driven approach to segmenting the policy arena" (Trauth 1986, p. 46), and instead suggests a systems approach "using the INPUT-PROCESS-OUTPUT (IPO) model as a basis for organizing policy contexts" (Ibid.). The virtue of this is that "orientation shifts to the view that information policy governs a *process* (such as the storage or transmission of information) rather than a *thing* (such as technology)" (Ibid.). But once describing this approach, Trauth cannot provide further clarification of how information policy should be evaluated. Furthermore, her research approach implicitly assumes an applied science paradigm for analysis that is reflected in her systems approach to this problem.

On the positive side, Trauth, like most writers on information policy, further assumes that her approach will improve information policies and make them more effective. This is, of course, the goal of all this evaluative activity. Another positive salient characteristic of information policy research is that investigators assume that the application of specific disciplinary knowledge to public policies will somehow produce desirable policies. This will be true to the extent that any discipline possesses theories that are able to prescribe actions that will eventuate in predictable results.

Recognizing this, the policy analyst is faced with at least three major problems. The first is determining which disciplinary knowledge is applicable to a given policy. In some cases, such as some policies regulating experimental drug testing, expert pharmaceutical knowledge is a primary, but not exclusive, ingredient in evaluation. But for other policies, especially those involving information, the decision regarding the application of disciplinary knowledge is not clearly evident nor is it known how such knowledge is to be applied. Many disciplines are probably involved in the evaluation of such policies. In fact, the contemporary definition of existing disciplines may further complicate the analyst's task. What the analyst is actually interested in is the knowledge obtained by researchers who are willy-nilly associated with one or more disciplines. The analyst's purpose is to identify relevant knowledge. The danger facing the analyst is that the application of inappropriate or insufficient disciplinary knowledge can result in flawed policies and may even create additional social and cultural problems.

A second problem requires much more honesty on the part of the analyst. That analyst must be able to determine whether the knowledge to be applied to the policy is of a kind that can predict the results of actions taken. For example, policy analysts know from experience that certain knowledge can *predict* what will happen under specific circumstances. However, they must not assume similar results when they apply this same knowledge under conditions where its predictive power has never been verified. Related, but different, knowledge may instead be required. In other words, predictability may be a condition for applicability. But analysts should not mistakenly place hope in the predictability of approaches that do not deserve it. Trauth unquestioningly assumes that the systems approach will lead to the truth in information policy analysis.

Finally, a third, practical problem consists of perfecting a methodology to solve the first two problems (which disciplinary knowledge and whether it is appropriate). One purpose of this study of the SATCOM report is to develop a reliable method for analyzing an information policy in order to decide what disciplinary knowledge is applicable for its evaluation.

SATCOM REPORT

In February, 1966, the National Academy of Sciences and the National Academy of Engineering established a Committee on Scientific and Technical Communication. The founding of this committee was in response to a request on October 20, 1965 by the National Science Foundation, which funded the work of the committee. Its general task was "to investigate the present status and future requirements of the scientific and engineering communities with respect to the flow and transfer of information" (*Scientific and Technical Communication*, 1969, p. iii).

SATCOM's work, which stretched over three years, was conducted by 25 members of the Committee itself, a staff of three, and over 200 consulting correspondents. It had three major objectives:

(1) "To gain a comprehensive overview of the current state and re-
    quired evolution of scientific and technical communication;"
(2) "To stimulate increased participation among individuals and insti-
    tutions in national planning for the improvement of scientific and
    technical communication;"
(3) "To function as a forum and clearinghouse on currently acute issues
    relevant to scientific and technical communication" (*Scientific and
    Technical Communication*, 1969, p. 285).

The Committee's report, over 300 pages long, consists of 10 chapters. Chapters 1 and 2 present general information about the committee and its *modus operandi* (Chapter 1) and a brief summary of its thinking about the subsequently presented recommendations (Chapter 2). Chapter 3, the core of the report, consists of 55 recommendations organized under the following rubrics: (a) Planning, Coordination, and Leadership at the National Level; (b) Consolidation and Reprocessing—Services for the User; (c) The Classical Services; (d) Personal Informal Communication; and (e) Studies, Research, and Experiments. Following each recommendation is a "discussion to indicate the thinking that led to its formulation as well as its action implications" (*Scientific and Technical Communication*, 1969, p. 5), that is, the assumed effect of the recommendation if it were carried out. Chapters 4–10 present additional detailed information on topics and problems covered in the recommendations.

The SATCOM report has engendered little critical comment. Within two years of its issuance only two articles did little more than summarize its content and point to tasks not addressed by it. Lesk (1970), for example, after summarizing the contents, remarked:

> The SATCOM Report is too vague and unspecific for its recommendations to produce very much progress. It discusses primarily the production of information, and makes useful suggestions for the writing of more and better review articles, increasing abstracting and indexing coverage, and reorganizing the financial structure and support of information distribution. Not enough attention is paid, however, to the currently severe problems of information selection and distribution, or to the introduction of promising mechanization and miniaturization techniques which offer the best hope of coping with the problems, providing economic relief, and improving the quality of information delivered to users. (p. 513)

Swanson (1970) made a similar evaluation. In spite of the deficiencies pointed out immediately after its issuance, the report seems to have stood the test of time well. Adkinson (1978, p. 129), for example, indicates a positive result of the report:

> Although no positive actions can be directly attributed to this committee's recommendations, the hearings it held and the discussions about them drew the attention of congressmen, federal administrators, industrial executives, and university administrators to urgent problems in sci-tech communication.

Finally, Aines (1984), 15 years after the report was issued, explains the failure of the major recommendation of the report in political terms. The report urged the formation of a Joint Commission on Scientific and Technical Communication, responsible to the National Academy of Sciences and the National Academy of Engineering. Aines remarked:

> The Commission was never formed, unfortunately. The two Academies did not have the funds to create such a Commission on their own. Hindsight tells us that it was a mistake not to keep the two Academies in the picture to provide a bridge to the scientists and engineers in the United States, the most important stakeholders in the evolution of science communications. (p. 181)

In summary, no detailed evaluation of the SATCOM report has taken place. What assessment has occurred, however, has either focused on what it did not do, on the political ramifications, or on the political reasons for its failure. The report has not been subjected to any structured disciplinary evaluation.

METHOD

The general method used in this case study is policy analysis. Policy analysis, however, includes many disparate techniques and approaches, many of which will not be employed here. Moreover, this study will only take the initial steps in performing a policy analysis. In order to clarify the approach taken here and its extent, imagine that a policy analyst has been assigned the task of analyzing a policy proposal. Because almost any policy proposal or implemented policy is a large entity, the analyst will first break it down into individual tasks. In order to do this it is assumed that each policy task has a goal and an actor or group of actors. Therefore, each policy task can be identified if individual goals and actors who are supposed to accomplish these goals are identified. Eventually the problem of determining whether all the individual policy task goals in the aggregate will achieve the overall goal of the policy will arise, but such a concern is not germane at this point. The major concern throughout the case study presented here is that the proper analytical tools be brought to bear in identifying and analyzing each policy task. Second, the analyst will try to identify the proximate (for definition see below) rationale associated with each task. Presumably, the rationale is closely bound up with the goal and actor of each task; this is the part of the policy task that can be subjected to disciplinary analysis.

Third, the analyst will determine what disciplines are applicable in evaluating the proximate rationale of each task. The analyst will then identify what discipline governed the rationale, determine the validity of the rationale, apply the appropriate disciplinary knowledge to each policy task, and eventually emerge from the process with an overall evaluation of the policy or policy proposal.

Method: General Comments

In this study, however, an attempt will only be made to ascertain whether the proximate rationale of each policy task can be evaluated by information science. In other words, it is assumed that when information science can evaluate the rationale, information science has developed the theories, models, empirical evidence, or analytical tools to determine if the rationale is legitimate and validated and if the goal of the policy will be reached. In this study analysis will cease after determining whether the rationale can be evaluated by information science.

The determination of the disciplinary boundaries of information science is also problematic. As the review of literature showed, the meaning of the term "information science" is not yet clear. For the purposes of this study, however, information science will be broadly defined as that discipline that deals with problems associated with controlling (the collection, preparation, analysis, translation, synthesis, storage, dissemination, access, or retrieval of) information or surrogates of such information such as abstracts, indexes, bibliographic citations, or reviews, or the collecting, preparing, storing, disseminating, and retrieving of information.

The case study consisted of a survey of people with experience in information science and information policy. They were asked several questions, some of which concerned the relation of information science to specific recommendations of the SATCOM report. Of course, there is no guarantee that those surveyed had the same notion of information science. The responses to the survey indicate roughly the extent to which a common mental model of information science was operative among those responding.

There is no evidence to indicate that this type of methodological analysis has been attempted before within information policy. Present methods of "analyzing" information policy do not indicate whether the policy proposal will be effective or whether the intended goals will be reached. Therefore, the results of this study could have a direct impact on the way information policy is analyzed, and lead to more effective information policy making.

PROCEDURE

This study has two parts. The first is a content analysis of the recommendations of the SATCOM Report. The second is a survey of experienced practitioners in the field.

Content Analysis

Content analysis is a broad term that has come to mean many different things to different researchers. Essentially it is a type of textual analysis, the aim of which is to identify the meaning of the text. Lindkvist (1981, p. 23) has described how the meaning of a text "can be identified with the producer, the consumer, and the interpreter of the text or with the text itself." In Lindkvist's formulation (1981, p. 34),

> Content analysis particularly consists of a division of the text into units of meaning and a quantificating of these units according to certain rules. . . . That

content analysis is systematic implies inclusion and exclusion of categories according to consistently applied rules. The possibility that the researcher will use only material supporting his hypothesis is thereby eliminated.

Andren (1981), in discussing the reliability of content analysis, further describes three different types of content analyses—pragmatic, semantic, and syntactic. The type of content analysis used in this study is pragmatic. A pragmatic content analysis focuses on the "intentions of the communicator, the goals of the communication" (p. 55).

This study assumes that the SATCOM recommendations were written to effect specific actions. Therefore, by identifying the actors, goal, and rationale of each recommendation, this study will be carrying out pragmatic content analysis, the purpose of which is to clarify the intention of the SATCOM Committee in each recommendation. Two studies have been identified that are similar, in some respects, to the present study.

Two Similar Studies

The first (Ellsworth, 1965) applied content analysis to the 1960 presidential campaign debates. He hypothesized that "in the 1960 Presidential Television Debates, the protagonists made clearer statements of their positions and offered more reasoning and evidence to support their positions, than they did in other campaign situations" (p. 794). To test his hypothesis he applied content analysis to selected portions of the debates in order to identify specific situations analyzed by the candidates, evidence for the position taken by the candidate, and a declaration of that position. He found that "the debating situation changed the type of discourse used by the candidates" (p. 802). The similarity of Ellsworth's study to the one undertaken here is the focus on the content analysis of discourse to identify the rationale from a specific position or stance.

In a different vein, and much more closely resembling the content analysis used here, Tutchings (1979) applied content analysis to over 100 reports of U.S. Presidential Commissions that appeared from 1945–1973. After training two coders and testing their reliability and consistency, he had them analyze the policy recommendations of these commissions. Among other things for each recommendation of these commissions, he identified the proposed actor, the type of action (according to a widely accepted policy typology), and the goal (or ends) of the proposed recommendation. Tutchings reported what he found about commissions in general over this period; he made no attempt, however, to identify the rationale for any of the recommendations. The way content analysis is applied in this study is not entirely novel, therefore, and has been used successfully in other studies.

Application of Content Analysis

This study proceeded as follows: Each of the 55 recommendations of the SATCOM report was subjected to content analysis. Each recommendation was first broken down into its constituent policy tasks by identifying goals and actors. (A policy task is something recommended by a policy that has a goal and an actor.) Rationale(s) for each task were identified. The rationales were classed into one of two groups: those that can be evaluated by information science and those that cannot. Tallies of the results of these identifications and classifications were made. Next, a group of experienced practitioners in the library and information science field were surveyed to find out how they would class a *randomly chosen subset of these rationales* . Then tallies of these identifications and classifications were made and analyzed. This survey was pretested on a group of 10 experienced practitioners in the field. Eight of the ten responded. After examining the responses, the same survey instrument was used without any change. Following are definitions of Goal, Actor, and Rationale, and the criteria for classifying rationales of policy tasks. A description of the survey is given and, finally, the assumptions and limitations of the study are presented.

Definitions of Policy Elements Identified
by Content Analysis

Policy Task. A policy task consists of a goal and an actor or group of actors to achieve the stated goal. Therefore, the identification of each policy task depends on identification of the goal and actor.

Goal. Any statement that presents the purpose or objective of the policy task. It may contain the words "should," "urge," "request," "must," "consider," etc.

> **Example:** "Must make a systematic effort to improve the quality and timeliness of formal publications." "Should assist and encourage its natural subdisciplinary groups to organize for and initiate the conduct of appropriate need-group services."

Actor(s). the person, institution, or group designated in the policy task to implement it. There may be more than one actor per policy task.

> **Example:** "Scientific and technical organizations and other publishers must make . . ." "Each larger scientific or technical society or association should . . ."

Rationale(s). A rationale is the fundamental reason for something, an exposition of principles or reasons, or a hypothesis. In the SATCOM report a rationale will be identified as any statement that utilized evidence to support the goal stated in the policy task:

- It may be true or false;
- It may cite consequences if carried out;
- It may cite consequences if not carried out; or
- It may not be exhaustive.

Example: "Lag times in publication of as much as a year must be considered intolerable. . . . [There is an] increasingly urgent need for prompt access to information about new developments." "It would greatly enhance the development and provision of such services." "Information must be transferred in usable form."

Rationale: Additional Comments

A common dictionary definition of the concept "rationale" seems straightforward: "The fundamental reasons, or rational bases, of something; or a statement, exposition, or explanation of principles." Often the term rationale has been used interchangeably with the word "theory." "Rationale" has been preferred, however, because "theory" carries with it connotations of scientific rigor or at least some notion of validity. A "rationale," on the other hand, can simply mean, as the above definition shows, a reason for doing something. The identification of the rationale for a policy task is important in this study because the rationale may be evaluated by information science or other disciplines. The identification of the rationale is not without its problems, however. In examining a policy task more than one rationale may be identified. In the broadest sense the rationale for any policy task of an information policy is to affect the the flow of information. The problem arises, however, in analyzing this broad rationale with the discipline of information science. A more narrowly perceived rationale, or proximate rationale, should be identified and subjected to analysis. If this is not done, then actions, such as those suggested by Recommendation C8 of the SATCOM report, can be considered as a rationale that can be evaluated by information science. Recommendation C8 from the SATCOM report deals with the payment of page charges for journal article publication. While an overarching (not proximate) rationale may be considered to fall within the criteria for classifying this rationale as information science—it deals with the dissemination of information through publication—its proximate rationale is financial. Therefore, the evaluation of the rationale cannot be carried out within the strict disciplinary boundaries of information science that are defined by the criteria. If an analyst were to evaluate such a policy task (which is not within the scope of this study, but well illustrates the legitimacy of this approach), the analyst would focus, to a significant degree, on the financial effect of such an action in the administration of research, its political impact, etc., all of which admittedly affect the dissemination of information. But the theoretical and practical disciplines that can assess the legitimacy of such actions are not those of information science. They may fall into other disciplines, such as management, accounting, and economics, which are often related to, but decidedly separate from information science.

In summary, while several rationales may be identified for a given policy task, in this study respondents were asked only to subject the proximate rationale to the criteria for rationale classification. The proximate rationale can be readily identified within the policy document by asking: "Why is this specific action recommended?" The answer most closely related to the recommended action is the rationale. Those surveyed were sent this explanation of rationale as well as the following criteria.

The rationale:

- Can be identified; and
- Deals with information science as you understand it.

They were also sent the broad definitional statement of information science that appeared earlier in this chapter.

Presentation of the Results of the Content Analysis

The actor, goal, and rationale are explicitly identified following the text of the recommendation. In many cases recommendations consist of more than one policy task. A rationale is identified and classified for each individual policy *task* within a recommendation. Furthermore, in order to minimize misidentification of the rationale, one of three possible sources is identified for each rationale: the recommendation itself, the SATCOM report excluding the recommendation, and inferential reasoning. The rationale is also classified according to the criteria presented above. In many cases, a discussion of the basis for the classification is also presented.

THE SURVEY OF EXPERIENCED PRACTITIONERS

The technique employed in this part of the study is the survey. Its purpose is to determine the reliability of both the content analysis of the recommendations of the SATCOM report and the subsequent classification of rationales into the information science or non-information science categories.

Sample

The emphasis of this survey is on the agreement of experienced practitioners about the applicability of information science to the evaluation of specific policy recommendations of the SATCOM report. It is reasonable to assume that experienced practitioners in the field will be able to determine if the discipline of information science can evaluate rationales for recommendations.

A sensible and reliable way to obtain confirmation of the method used, and one that was used in this study, is to identify 45 individuals who were on the SATCOM panel, who served as consulting correspondents for SATCOM, or who engaged in research related to this topic. To be included in the latter class, a person must have written on information policy or on the definition, nature, or content of the discipline of information science during the five year period 1981–1986. They were contacted personally by the investigator. These individuals were deliberately chosen for their noted expertise. The choice was not random. They were chosen because they were qualified to respond, and were likely to be interested in this research. Of course, not all those requested to participate did so. Twenty responded. This number of respondents is comparable to those used in Delphi studies, a method to which this survey is similar. The individuals selected were asked to read 15 randomly selected recommendations and the statements of rationale and to indicate for each task if the discipline of information science, as they understood it, could be used to evaluate the rationale for that task. These respondents were then sent copies of the recommendations and their rationales as identified by the investigator. By seeking confirmation of these experts, a reasonable indication of the reliability of the method was obtained.

Although these responses are subjective, the opinions of a body of researchers who have worked in this area are operationally authoritative. Furthermore, this survey serves only as a confirmation of the investigator's research method and of the analysis of the recommendations. This way of confirming a judgment by a group of experts is similar to, but not coincident with, the Delphi Technique (Linstone and Turoff, 1975). Such a method is relevant when, among other things, "the problem does not lend itself to precise analytical techniques but can benefit from subjective judgments on a collective basis" (Ibid. 1975, p. 4).

Data Collection

Each person was sent a questionnaire consisting of a cover letter, instruction sheet with examples, and a data collection form. The form actually consisted of several pages. Each page had a recommendation, an explanation of the recommendation (i.e., an identification of its goal and the proposed actors) and a statement of the rationale for each task of the recommendation. The respondent was asked to answer two questions. The first asked if the respondent agreed or disagreed with the content analysis of the task. The second asked if information science, as the respondent understood it, was suitable for evaluating the stated rationale, or whether some other discipline would better be able to evaluate it. A comments section for each task was also provided.

Pretest

This survey instrument was pretested on ten experienced practitioners using five randomly chosen recommendations from the Report. Eight responded. After examining the responses, the investigator was satisfied as to the instrument's efficacy. The same instrument was used with the survey group of 45 experienced practitioners. However, 15 different randomly chosen recommendations were used.

Data Analysis

Data were tabulated and analyzed manually. Generally speaking, frequency distributions were computed on the data. The logic of data analysis, however, requires some explanation.

First, frequency tables for the first question (agreement or disagreement with the rationale) were computed. Agreement among respondents to each individual recommendation was determined. A scale was used to describe the degree of agreement or disagreement. If 70–100% of the respondents agree or disagree with the stated rationale then this shall signify consensus; if less than 70% agree or disagree then this shall signify nonconsensus. Summaries of the number of recommendations that fall into each of these categories, as well as explanations offered for the emergent aggregate patterns, and for specific recommendations that warrant such explanation will be provided in Chapter 8. Second, responses to the second question (classification as information science or not) were also tabulated and dealt with in a similar manner. The results of these analyses are presented in the subsequent chapters.

ASSUMPTIONS OF THE CASE STUDY

SATCOM Report

SATCOM recommendations were written to effect specific actions. This is a cornerstone of much of the analysis. If a recommendation were not written to effect a specific action, then, one might ask, why was the recommendation made at all? Note carefully that this assumption does not specify what type of action is intended.

The rationale for a given policy task, if it exists, is contained in the SATCOM recommendation, the report itself, or can be inferred by the investigator. This assumption implicitly *suggests* a rationalist model for policy making. By making such an assumption, however, the investigator by no means excludes other types of models. It simply means that a recommendation is made because of a reason. The assumption only states that there is a reason for the recommendation, not necessarily that it is a rational reason.

Method

Criteria can be written to identify actor and goal for a policy task rationale. This is assumed based on prior experience with content analysis and goes back to the basic premise of content analysis itself as a valid method for social science research. If criteria could not be written, the reliability of the method for consistently identifying actor, goal, and rationale would be suspect.

Individuals surveyed were qualified to respond. If they were not qualified then their responses would not mean much in the context of this study. But the individuals surveyed were chosen based either on their participation in SATCOM or on their published works, current position, or participation in professional activities associated with information science.

Identifiable knowledge does exist that can evaluate a rationale. The investigator has tentatively identified some aspects of this knowledge by naming economics, sociology, law, political science, and information science.

Information Science

It is assumed that information science is a body of knowledge that contains methods and procedures that can evaluate rationales of information policy. If it were shown, as some have tried to do, that information science is neither a science nor a discipline, this would not affect those methods and procedures commonly associated with information science. They could still be used to evaluate a rationale. If a rationale can be evaluated, some body of knowledge must be able to do it. This study assumes that information science is part of that body of knowledge. If information science does not exist, as some have maintained, it in no way eliminates that previously identified body of knowledge that can evaluate the rationale. One could simply not identify that knowledge as information science.

Information Policy

The results of this study will be useful in improving the analysis of STINFO policy and creating more effective STINFO policy. It is intended to develop a model procedure that can be applied to other information policy documents. Of course, the validity of this assumption may not be tested until it is applied to other policies.

STUDY LIMITATIONS

The results of this study alone will not be sufficient to evaluate STINFO policy. Several other analytical tools, based on different disciplines, may be required. Furthermore, the application of this method to other types of information policy documents may not be effective.

CHAPTER 7

EVALUATION OF SELECTED SATCOM RECOMMENDATIONS

The following randomly selected SATCOM recommendations are arranged as they were in the SATCOM report. Each recommendation is followed by a section labeled "Analysis of Recommendation," which consists of the content analysis of the recommendation and an indication of the source where the rationale was found. A page reference in parentheses will follow any part of the rationale taken from the report itself.

This "Analysis of Recommendation" is followed by three other items dealing with the survey of experienced practitioners. The first additional item is a tabular description of the responses received. This is divided into two parts, corresponding to the two questions asked of the respondents. Part one indicates the number and percentage of the respondents who agreed or disagreed with the adequacy of the content analysis prepared by the investigator. The second part indicates the number and percentage of the respondents who agreed or disagreed that the discipline of information science could evaluate the validity of that recommendation's rationale.

The second additional item consists of an analysis of the responses received from those surveyed for that recommendation. This is also divided into two parts, one discussing the adequacy of the content analysis question and the other discussing the ability of information science to evaluate the rationale's validity. Consensus is indicated if the percentage agreeing falls within the ranges 0–30% or 70–100%. If the percentage is in the 0–30% range, this indicates a consensus of disagreement with the question asked (i.e., adequacy of content analysis or ability of information science to evaluate the rationale). If the percentage is in the 70–100% range, this indicates a consensus of agreement with the question asked.

The third additional item consists of the investigator's comments and analysis of the responses received. The remainder of the SATCOM recommendations, as well as an example of the survey instrument, can be found in Burger (1988). Only content analysis and an indication of the source of the rationale are provided for the recommendations given there.

ANALYSIS OF RECOMMENDATIONS

A. Planning, Coordination, and Leadership at the National Level[1]

A2. Effectiveness and economy demand a basic philosophy of shared responsibility between private organizations—those for profit and those not for profit—and the federal government in the management of scientific and technical information. In this sharing, the major scientific and technical communities and organizations involved in major information-handling activities should exercise leadership in improvement and management, recognizing the place of their activities as part of a national aggregate of endeavor in which the government also plays a major role. Equally, all government agencies should rely on organizations of the relevant scientific, technical, and information-handling communities for a major share in the management of the information services required by agency missions and activities.

Analysis of Recommendation

Actor(s): Public and private organizations that deal with scientific and technical information.

Goal: Improved management of scientific and technical information and realization by the actors that public and private organizations each play important roles in the management of STINFO.

Rationale: There was a perceived need for better management of STINFO activities. There was a perceived need for both private and public organizations to recognize each others' role in this goal. It was recognized that both types of organizations had much to offer each other.

Source of Rationale: Recommendation.

---

[1]These bold-faced headings are taken directly from the SATCOM report. The SATCOM recommendations have been italicized for easier identification.

Respondents—Survey Proper

	Agree	Disagree	No response
Adequacy of content analysis:	19(95%)	1(5%)	
Information science can evaluate			
rationale's validity:	11(55%)	9(45%)	

Survey Analysis

Adequacy of Content Analysis: The one respondent who disagreed with the content analysis stated that the Federal government was also an actor, and this was not specifically mentioned in the content analysis.

Ability of Information Science to Evaluate the Rationale's Validity: According to the rating scale there was a lack of consensus on this item. Several of the respondents made comments regarding their choice. One respondent based his disagreement on the fact that private organizations are often profit-oriented "with many stockholders, investors, etc. Marketing products for profit is not within information science discipline." Another stated that "these are economic/political issues outside the 'discipline' but of concern to its members." One who agreed replied: "The individual information scientist may not be fully aware of the agencies' 'missions and activities.' Leadership should be exercised by the responsible agency with advice from information science."

Comments: In all of these comments two observations are in order. First, the respondents indicate that the issues raised bear some relationship to information science, as well as to other disciplines. This indicates that in analyzing the rationale some combination of information science and other disciplines must somehow come together to make a judgment. Second, the type of 'advice' to be given, or the issues that are "of concern to its members" indicate a definite interaction with the social order. This suggests that the advice given and the issues raised have definite social implications and go beyond a 'value-free' scientific discipline. The lack of consensus may be attributed to the fact that the respondents failed to distinguish between the ultimate and proximate rationales. The ultimate rationale for this suggestion was aimed at improving the scientific and technical information system and was explicitly stated in the goal. The proximate rationale was the expected beneficial effect of greater cooperation between the private and public sectors, an action that would hopefully achieve the ultimate goal.

B. Consolidation and Reprocessing—
Services for the User

B5. Each society or association, the membership of which includes many persons concerned with practice, especially in engineering, medicine, and agriculture, should increase substantially its attention to information programs that will:

1. Ensure that access, awareness, and appraisal services comparable to those supplied for the body of research literature are provided also for publications of particular interest to the practitioner, such as textbooks, monographs, handbooks, manuals, patents, trade journal publications, company reports, catalogs, specifications, and standards.

2. Stimulate the production of critical reviews and surveys of contemporary fields of knowledge, the condensation being focused on particular domains of application of interest to the practitioner and adapted to his needs.

3. Identify types of data banks, including diverse types such as Sweet's Catalog, the Chemical Compound Registry (of Chemical Abstracts Service), and the Thermophysical Properties Research Center at Purdue, which need to be established in a field; establish or foster the creation of required data banks; and provide an indexed inventory for existing ones that describes coverage and conditions of access.

4. Meet the needs resulting from requirements of continued education to keep practitioners in its field up to date.

Analysis of Recommendation

Actor(s):   Societies and associations of applied science.

Goal: Create information programs that will make public documentation easily available to practitioners of applied science.

Rationale: "Documentation, especially public documentation, of new ideas, is given much less attention by technologists than by scientists" (p. 43). "Convenient access to and reprocessing and repackaging of these various kinds of technical information are sorely needed" (p. 44). Societies and associations are in the best position to create/foster these information programs.

Source of Rationale: Report

Respondents—Survey Proper

	Agree	Disagree	No response
Adequacy of content analysis:	15(75%)	5(25%)	
Information science can evaluate rationale's validity:	14(70%)	6(30%)	

Survey Analysis

Adequacy of Content Analysis: Although five respondents disagreed with the content analysis, there was consensus on the adequacy of the content analysis. One specifically questioned the omission of individual members of the associations mentioned; another criticized the omission of the requirement to evaluate the public information made available through these programs.

Ability of Information Science to Evaluate the Rationale's Validity: The responses indicated consensus on this recommendation. Those agreeing with the ability of information science to be applicable here made the following comments: "This is a need determination problem and information science has a depth of needs analysis techniques to assert the validity of the rationale"; "bibliometric techniques could verify validity"; and "information science can (and should) certainly work toward a determination of what information professionals do and what user groups do." Those disagreeing seemed ambivalent. One, while disagreeing, stated: "Part of this, *re* the degree of technology documentation, can be studied by information science." Another also appeared to contradict his disagreeing response: "User clients, e.g., doctors and health professionals, and their requirements for information can be analyzed by information scientists, but information scientists have limited role in changing their work habits or information-seeking behaviors."

Comments: Again information science is perceived as a multidisciplinary field that has access to many analytical techniques. The ambivalence in the respondents' comments may be seen as a reflection of this perception.

B10. These societies and agencies concerned with the conduct and support of abstracting services should seek actively to identify difficulties, find solutions, and take the initiative in proposing and testing arrangements through which an increasing contribution by the sponsors of research to the input costs of the basic abstracting services can make transfer for reprocessing financially possible at approximately output costs.

Analysis of Recommendation

Actor(s):   Societies and agencies concerned with the conduct and support of abstracting services.

Goal: Actor should evaluate present abstracting services and devise ways that sponsors of research can contribute to the input costs of basic abstracting services so that these services can make their product available for reprocessing at a lower cost.

Rationale: Repackaging and reprocessing of abstracts are necessary for scientific communication. Sponsors of research should contribute to the input costs of these services because support of research does not stop with the publication of results. The actor is in the best position to determine the arrangements through which this support should take place.

Source of Rationale: Report and inference.

Respondents—Survey Proper

	Agree	Disagree	No response
Adequacy of content analysis:	18(90%)	2(10%)	
Information science can evaluate rationale's validity:	13(65%)	7(35%)	

Survey Analysis

Adequacy of Content Analysis: The responses indicated a consensus on the adequacy of the content analysis. Those disagreeing with the content analysis made no comments.

Ability of Information Science to Evaluate the Rationale's Validity: The responses to the second question indicated a lack of consensus. One disagreeing respondent stated: "It seems to me that economics of information draw [sic] heavily on economics, not information science as the mother lode or discipline." Another stated: " 'Should contribute' is a moral imperative (morals are human not disciplinary)." Those agreeing clarified their responses by these remarks: "Basic research in abstracting and indexing is carried out by information science faculties in universities. Not many abstracting and indexing services have the staff or the capability to find new solutions for improving scientific communication." Another said: "This is a cost/benefits problem and so [the] information science discipline has been concerned with this for quite a few years now." And again: "Abstracting and indexing plus vocabulary control are within information science. Cost analysis may slide into operation research."

Comments: The lack of consensus may be explained by the ambivalence of the respondents in deciding whether cost accounting, or any type of economics, belongs to the domain of information science.

B14. The Commission (proposed in Recommendation A1) should assist in the identification of major information analysis centers that are operating in particular subject areas and have the capability of offering services fulfilling need-group requirements in these areas. Further, the Commission should stimulate and aid in the exploration of ways in which such services can be made more widely available.

Analysis of Recommendation

Actor(s):   Commission.

Goal: (1) assist in identifying IACs in specific subject areas; and 2) help explore ways to make these services more widely available.

Rationale: "While present experience seems to limit the intellectual and economic viability of such centers to certain rather specific fields, it is clear that their potential is still far from fully exploited. The pricing of services provided by the centers requires careful consideration to ensure both economic viability and wide use" (p. 50).

Source of Rationale: Report.

Respondents—Survey Proper

	Agree	Disagree	No response
Adequacy of content analysis:	18(90%)	2(10%)	
Information science can evaluate rationale's validity:	11(55%)	9(45%)	

Survey Analysis

Adequacy of Content Analysis: Most respondents agreed that the content analysis was adequate. Those who disagreed made no comments.

Ability of Information Science to Evaluate the Rationale's Validity: There was a lack of consensus on this question. One respondent did agree that "the Commission has the proper staff of information scientists and the resources." Another stated: "Again this is another needs analysis problem." A third responded: "Analysis of subject strengths and needs of subject specializations is in [the] domain of information science." Those disagreeing also made comments. One said: "The users of these centers belong to disciplines *other than* information science." Another replied: "In this case the rationale seems vacuous." A third warned: "Joint judgment of information science and the scientific community [is] essential." Another stated: "The validity of information analysis centers must come from the user disciplines, not information science."

Comments: The absence of consensus here may be explained by the lack of clarity in the wording of the quoted rationale. On the other hand, the inability of any respondent to identify clearly the domain of information science versus other disciplines to which it is related could legitimately be seen as a reason for the absence of consensus.

C. The Classical Services

C4. The National Federation of Science Abstracting and Indexing Services should seek support for the development and promulgation of guidelines to be followed by editors and publishers of primary information in specifying the required documentation units [i.e. author supplied abstracts, index terms, etc.].

Analysis of Recommendation

Actor(s): National Federation of Science Abstracting and Indexing Services.

Goal: Seek support for development and promulgation of guidelines for author-supplied documentation units.

Rationale: Other bodies have promulgated guidelines for the preparation of documentation units. "By eliminating some of the more obvious and frequent incompatibilities, the recommended guidelines should remove a major obstacle to the acceptance of the documentation-unit concept" (p. 58). This will allow recommendation C3 to be carried out and achieve its goal.

Source of Rationale: Report.

Respondents—Survey Proper

	Agree	Disagree	No response
Adequacy of content analysis:	20(100%)	0(0%)	
Information science can evaluate rationale's validity:	14(70%)	6(30%)	

Survey Analysis

Adequacy of Content Analysis: Every respondent agreed with the adequacy of the content analysis. One respondent warned: "I partly agree but I believe that ANSI Z39[2] standards apply in this area and that perhaps a combined approach of the National Federation of Science Abstracting and Indexing Services and ANSI Z39 as actors is needed."

Ability of Information Science to Evaluate the Rationale's Validity: There was a consensus regarding the ability of information science to evaluate the validity of the rationale. One agreeing respondent stated: "Sounds like bibliographic control work to me and if we can't claim that we may as well give up." Another offered a different perspective on the problem: "Every one of these recommendations' rationales has an aspect—namely *who* 'should' do something—the *validity* of which cannot be *determined* by any discipline; it is governed (not determined) by a political process. In a sense this *has* its own 'discipline' but it is not a *discipline* in the sense of a scientific discipline." One respondent who disagreed stated: "Study of 'acceptance' of an idea sounds like sociology of innovation."

Comments: In spite of the consensus, the comments indicated ambivalence. This was probably occasioned by two things. First, the distinction between the actor and the action gave rise to one respondent's doubt. Second, information science would not easily be separated from another discipline or body of knowledge working in conjunction with it.

C6. The U.S. Office of Education should support a broad program in library education (in addition to or as part of its present program under Title IIB of the Higher Education Act), with special attention to the following objectives:

1. The training of more students in systems analysis, systems planning, and operational analysis of libraries and library services.

2. The training of all students as well as faculty (throughout their college careers if feasible) in the use of the increasingly complex array of existing library and information services.

Analysis of Recommendation

Actor(s): U.S. Office of Education.

Goal: Support broad program in library education.

---

[2]The American National Standards Institute (ANSI) promulgates standards in a variety of areas, including matters relating to libraries and information transmission. Z39 is the committee of ANSI responsible for library and information transmission standards.

Rationale: "If the application of modern technology seems to lag in libraries as compared with other areas, it is because the scope of the problem is immense, and more particularly, because our society and its institutions are not effectively organized for a direct attack on the broad systems-planning problems that lie at the heart of the matter" (p. 59–60). Two ways to mount this attack are the training of library students in library systems analysis and training of all students in the use of library and information services.

Source of Rationale: Report and inference.

Respondents—Survey Proper

	Agree	Disagree	No response
Adequacy of content analysis:	16(84.2%)	3(15.8%)	1
Information science can evaluate rationale's validity:	15(78.9%)	4(21.1%)	1

Survey Analysis

Adequacy of Content Analysis: There was a consensus regarding the adequacy of the content analysis.

Ability of Information Science to Evaluate the Rationale's Validity: There was a consensus that information science could evaluate the validity of the rationale. One respondent commented: "BI [bibliographic instruction] evaluation techniques could be applied though these are, after all, education techniques applied to BI." A final agreeing respondent replied, "Each of these recommendations is normative, asserting what should be done. The validity of a rationale would be more easily justified if it can be formulated in if–then terms., e.g., if the US Office of Education were to support a broad program in library education, then . . . then, the rationales in most cases would give necessary but not sufficient antecedents." One who disagreed stated that "'explaining [indecipherable words] is not a problem addressed by a discipline. Perhaps an historian or 'policy analyst' could help." Another said "USOE is a 'populist' educational agency. It is concerned with school children, college students, etc. Information scientists, for the most part, have not concerned themselves with the information needs, behavior, of such groups."

Comments: Even with consensus on this recommendation, some respondents admitted an ethical dimension in the practice of the discipline of information science.

C7. We recognize the need for systematic analysis and study of the economic aspects of formal scientific and technical publications over the next five to ten years. Such a study should examine the income returned to such publications from their principal markets—users, authors, and the public—together with trends in cost factors and the impact of new technologies, to serve as a basis for the development of flexible funding and pricing policies, which, in a changing environment, should be responsive to the needs of each interested party without being unduly responsive to any.

Analysis of Recommendation

Actor(s):   Unspecified.

Goal: Systematic analysis and study of the economic aspects of formal scientific and technical publications.

Rationale: "Since publication costs have constituted a perennial problem, especially for the not-for-profit scientific and technical organizations, a variety of economic mechanisms have been devised to meet these costs and have resulted in a complex pattern of interdependencies" (p. 65). Such study is needed "as a basis for the development of flexible funding and pricing policies" (recommendation).

Source of Rationale: Report and recommendation.

Respondents—Survey Proper

	Agree	Disagree	No response
Adequacy of content analysis:	16(84.2%)	3(15.8%)	1
Information science can evaluate rationale's validity:	11(55%)	9(45%)	

Survey Analysis

Adequacy of Content Analysis: There was a consensus about the adequacy of the content analysis. Of the three who disagreed, only two made comments regarding the reason for this disagreement. One stated that the Commission is the intellectual actor. The other stated that the actors were students and investigators of the economic aspects of formal and informal scientific communication.

Ability of Information Science to Evaluate the Rationale's Validity:
There was no consensus concerning the ability of information science to evaluate the rationale. Several respondents made comments. Only one who agreed did so. He said that other disciplines, such as economics, would also be applicable. One respondent who disagreed said that "clearly an econometric model is required (also a Marxist point of view)." Another stated: "Porat, Machlup, et al. come from economics. They could not be called, in my book, information scientists." A third simply said: "Sound [sic] like we need an economist here."

Comments: The main reason for nonconsensus seems to be a disagreement over the relation between information science and economics.

C10. Major scientific and technical societies (if not already doing so) should experiment with:

   1. A journal for brief, refereed, and promptly published papers (letter journal), with issue period and publication lag not exceeding one month.
   2. Organized reprinting, from a group of journals covering either a narrow area or a group of cognate fields, of selected papers recognized as most outstanding.

Analysis of Recommendation

Actor(s):   Major scientific and technical societies.

Goal: Experiment with different means of publishing research results.

Rationale: Reducing lag times is a worthwhile goal. Some research published as technical reports is not made widely available. Such experimentation as recommended would determine whether these solutions would be successful.

Source of Rationale: Report and inference.

Respondents—Survey Proper

	Agree	Disagree	No response
Adequacy of content analysis:	17(89.5%)	2(10.5%)	1
Information science can evaluate rationale's validity:	14(73.7%)	5(26.3%)	1

Survey Analysis

Adequacy of Content Analysis: Only two respondents disagreed with the adequacy of the content analysis. One made comments; the other did not. The one who did, however, simply indicated that a phrase should have been added to the recommendation, but did not criticize any part of the content analysis.

Ability of Information Science to Evaluate the Rationale's Validity: There was a consensus on the ability of information science to evaluate the rationale. Of those who agreed, several made comments. One said, "Bibliometrics could test validity." Although this respondent did not circle an answer for this question, his response indicates agreement. He said, "Information science can determine whether lag time and lack of availability exist, scholarly societies and organizations and publishers can initiate experimentation." Those disagreeing with this question also commented. One asked rhetorically: "How would information science, a hybrid drawn from many disciplines, help an engineer or physician to decide what papers are outstanding in their respective fields?" Another playfully retorted: "This is getting warmer since information science studies questions of how widely items are disseminated, but the determination of 'worthwhile goals' is outside the discipline." A third replied: "Initial publication rests with the originator community which is also the ultimate user community."

Comments: In spite of the consensus, ambivalence is clearly evidenced by the respondents' comments. Still, some try to divorce the discipline from ethical considerations (e.g., the comment about "worthwhile goals," above).

C11. Scientific and technical societies should recognize the need for publishing information that facilitates informal scientific and technical communication in their fields (sometimes referred to as meta information). Information on "who is doing what and where" can appear in newsletters, as supplements to substantive journals, or as separate publications. This support of interpersonal and intraorganizational communication should be but one approach in a continuing program to facilitate information exchange.

Analysis of Recommendation

Actor(s):   Scientific and technical societies.

Goal: Publish meta information.

Rationale: Meta information facilitates information exchange.

Source of Rationale: Recommendation.

<u>Respondents—Survey Proper</u>

	Agree	Disagree	No response
Adequacy of content analysis:	17(94.4%)	1(5.6%)	2
Information science can evaluate rationale's validity:	15(78.9%)	4(21.1%)	1

<u>Survey Analysis</u>

Adequacy of Content Analysis: There was consensus concerning the adequacy of the content analysis.

Ability of Information Science to Evaluate the Rationale's Validity: The responses indicated consensus concerning the rationale for this recommendation. One respondent who agreed stated: "This is understanding scientific communication again. Though somewhat sociological it is taught in information science. INFROSS[3] study tried to do this." One who disagreed commented: "Information science does not have anything to do with this as it appears that meta information is merely 'trade or profession news' and is a journalistic problem." Another ambivalently stated: "Again, information science can assist in determining validity." A third suggested looking at his response to C10 in which he stated: "Initial publication rests with the originator community which is also the ultimate user community."

Comments: Again, comments suggest that information science is to be used in concert with other disciplines in order to evaluate the rationale.

D. Personal Informal Communication.

D2. Employers of scientists and technologists—universities, industrial and government laboratories, and others—should provide systematic opportunities, through leave and exchange arrangements, for long-term or summer visits to other institutions by staff members likely to benefit from them and, in turn, should welcome such visitors from other institutions. There should be general acceptance of this responsibility. In addition, the funding of fellowship programs should take into account the need for the continued generation of new contacts by mature professionals as well as younger ones. Special attention should be given to the needs of workers in small or out-of-the-way institutions.

---

[3]The INFROSS (Investigation into the Information Requirements of the Social Sciences) study was conducted in 1969–1970.

Analysis of Recommendation

Actor(s):   Employers of scientists and technologists.

Goal: Provide opportunities for interinstitutional visits of scientists and technologists.

Rationale: "The major role played by informal, person-to-person communication in the dissemination of information is generally recognized and has been confirmed in systematic studies of communication channels" (p. 75). "We believe that one aspect of interpersonal communication that can be significantly influenced by organization policies is the development of close personal contacts between workers in different organizations" (p. 77).

Source of Rationale: Report.

Respondents—Survey Proper

	Agree	Disagree	No response
Adequacy of content analysis:	16(88.9%)	2(11.1%)	2
Information science can evaluate			
rationale's validity:	7(38.9%)	11(61.1%)	2

Survey Analysis

Adequacy of Content Analysis: All but two of the respondents agreed with the adequacy of the content analysis.

Ability of Information Science to Evaluate the Rationale's Validity: There was no consensus regarding the ability of information science to evaluate the rationale. The majority of respondents, however, disagreed that it could. One respondent who disagreed stated: "Although the idea is a good one, information science has no special competence in determining the validity of a fellowship program." Another said: "This problem concerns all scientific disciplines so it is not purely an information science problem." A third replied: "Interpersonal contact will always exist in addition to (and will probably be preferred to) information science activities. Our role begins in the recognition that this process will always be inadequate." One respondent who did not circle a response stated, "That an organizational policy can have a certain effect is the concern of management science, perhaps. Information science can, of course, study the degree of person–person communication, but so can 'communication research' and so can sociology." Finally, a respondent who agreed claimed that "insofar as studies of scientific communication are part of information science, information science does provide techniques."

Comments: Lack of consensus may be attributed to the perception, evidenced by the comments, that many disciplines may be applicable here in addition to information science.

E. Studies, Research, and Experiments

E2. Among the principal objectives of the previously recommended studies (Recommendation E1) should be the development of measures of value for information services that embody various combinations of accuracy, completeness, discrimination, timeliness, and similar factors. Where appropriate, these studies should include experiments on user response to new services. (See also Recommendation E4.) The facility of providing additional or specialized information services at appropriately scaled prices should receive particular attention.

Analysis of Recommendation

Actor(s):   Unnamed.

Goal: Development of measures of value for information services.

Rationale: "Such studies will provide a better foundation for monitoring the performance of information programs as advocated in Recommendation B15"(p. 80). "Definitive system studies are clearly needed and should include all aspects of retrieval, handling time, and costs" (p. 80).

Source of Rationale: Report.

Respondents—Survey Proper

	Agree	Disagree	No response
Adequacy of content analysis:	15(83.3%)	3(16.7%)	2
Information science can evaluate rationale's validity:	17(89.5%)	2(10.5%)	1

Survey Analysis

Adequacy of Content Analysis: There was a consensus regarding the adequacy of the content analysis. Three respondents disagreed. One of these stated, "Although E2 does not name an actor, I believe that the responsible actors should be information scientists-researchers in the universities." Another, in the same vein, said: "Excludes implied actor."

Ability of Information Science to Evaluate the Rationale's Validity: There was a consensus regarding this question. Of those agreeing, one said: "Information science is concerned with studies of information systems and services so it should be able to support this rationale." One who disagreed said, "Who will determine 'better' and what is 'clearly needed' —not information science, sounds like market research." Finally, another disagreeing respondent said: "Again, economics is not a unique preserve of information science."

Comments: Apparently many respondents perceived information economics to be within the domain of information science in this particular case. This perception did not seem to hold true in other cases, however. (See, for example, the responses to recommendations B10 and C7.)

E5. Continued experimentation in the design and use of effective combinations of machine and human functions, both in preparing for and conducting searches for information, should be supported. Such combinations very possibly may not involve indexes of any conventional sort.

Analysis of Recommendation

Actor(s): Researchers.

Goal: "Continued experimentation in the design and use of effective combinations of machine and human functions" in information retrieval.

Rationale: Applications of advanced technologies may increase the usefulness of conventional retrieval tools.

Source of Rationale: Report and inference.

Respondents—Survey Proper

	Agree	Disagree	No response
Adequacy of content analysis:	13(72.2%)	5(27.8%)	2
Information science can evaluate rationale's validity:	18(100%)	0(0%)	2

Survey Analysis

Adequacy of Content Analysis: There was a consensus regarding the adequacy of the content analysis, but five respondents disagreed. None made comments regarding this question. One respondent, however, stated, apparently in reference to the recommendation itself: "Partial disagreement. The key word here is 'support.' Researchers can do the work; support must be provided by the Commission or NSF."

Ability of Information Science to Evaluate the Rationale's Validity: There was a consensus regarding the ability of information science to evaluate the rationale. One respondent said: "Information science is concerned with retrieval of knowledge and thus with its organization so thus [the] rationale." Another stated "Information storage and retrieval provide methods." Another said: "If 'useful' is defined in a pragmatic way, information science can study this." One respondent who did not answer the question stated, in apparent disagreement: "Most of the personnel in our computer science department are graduate engineers, and I think they regard engineering as their parent discipline, not information science."

Comments: None.

E7. The scientific and technical societies must use their information and publication programs to familiarize their members with experiments that explore the uses of advanced technology as a working tool in the communication of scientific and technical information, rather than just in support of documentation functions, and must insist on the participation of scientists, engineers, and practitioners with proven discipline competence in guiding and evaluating such experiments.

Analysis of Recommendation

Actor(s):   Scientific and technical societies.

Goal: "Familiarize their members with experiments that explore the uses of advanced technology as a working tool in the communication of scientific and technical information," and employ experts in guiding and evaluating such experiments.

Rationale: (Goal 1) Structures for "facilitating intercommunication between computer-processable files of diverse origins . . . give rise to problems that are predominantly of a basic conceptual nature rather than problems of equipment and implementation" (p. 83). The familiarization of members with these experiments may help solve these conceptual problems.

Source of Rationale: Report and inference.

Respondents—Survey Proper

	Agree	Disagree	No response
Adequacy of content analysis:	17(94.4%)	1(5.6%)	2
Information science can evaluate rationale's validity:	13(68.4%)	6(31.6%)	1

Survey Analysis

Adequacy of Content Analysis.   There was consensus on the adequacy of the content analysis. One who disagreed made no comments.

Ability of Information Science to Evaluate the Rationale's Validity: There was not consensus regarding the ability of information science to evaluate the rationale. One who agreed stated: "IS [Information science] probably is valid—because [indecipherable words] can do most of this better. The SATCOM report, like others, was endlessly fascinated by end-user information responsibility, a concept promulgated by Alvin Weinberg." Another stated: "Partial agreement. Information science will need to draw on the knowledge of other disciplines—cognitive science, telecommunications, displays, etc." One who disagreed said: "Sociology [should be applied] to convince [the] field [of information science] unless [there is] bibliometric proof."

Comments: Lack of consensus may possibly be attributed to the perception, evidenced by the comments, that information science must be used along with other disciplines.

Rationale: (Goal 2) Experts who know the problems associated in using advanced technology to facilitate communication of scientific and technical information are in the best position to guide these experiments.

Source of Rationale: Recommendation and inference.

Respondents—Survey Proper

	Agree	Disagree	No response
Adequacy of content analysis:	16(88.9%)	2(11.1%)	2
Information science can evaluate			
rationale's validity:	13(68.4%)	6(31.6%)	1

Survey Analysis

Adequacy of Content Analysis: There was consensus regarding the adequacy of the content analysis. One who disagreed commented: "Goal 2 implies *subject* experts (e.g., physicists and chemists) but analysis implies information scientists, not necessarily with subject knowledge."

Ability of Information Science to Evaluate Rationale's Validity: There was no consensus on this question. One who disagreed stated: "Partial disagreement. Information science has some, but not all, expertise in this area." Another flatly declared: "Not a researchable question, in my view." One who agreed said: "Analysis of results could be conducted using controlled experiment groups with bibliometric applications." Another reasoned: "Yes, *using* is more important than designing (computer science). The designers need specifications to work to."

Comments: Lack of consensus is attributed to the same reasons cited under Goal 1 of this recommendation.

E8. On-going activities directed toward the development and evaluation of languages for describing the formats of files as well as of alphanumeric and other digital communication are of immediate and key importance to scientific and technical communication. Thus, the National Science Foundation should cooperate with other federal agencies pursuing active programs in this area, especially the Advanced Research Projects Agency, to ensure rapid and coordinated progress. In particular, an evaluation program for a file-format language should demonstrate computer conversion from one format to another as soon as possible and for an extensive set of samples.

Analysis of Recommendation

Actor(s): National Science Foundation and other Federal agencies.

Goal: Progress in developing and evaluating file-format languages.

Rationale: Languages describing formats of files and other digital communication are important for scientific–technical communication. Rapid progress can be best attained by cooperation among NSF and other Federal agencies.

Source of Rationale: Recommendation.

Respondents—Survey Proper

	Agree	Disagree	No response
Adequacy of content analysis:	15(83.3%)	3(16.7%)	2
Information science can evaluate rationale's validity:	13(72.2%)	5(27.8%)	2

Survey Analysis

Adequacy of Content Analysis: There was consensus regarding the adequacy of the content analysis. Those who disagreed made no comments regarding the reason why.

Ability of Information Science to Evaluate the Rationale's Validity: There was a consensus regarding the ability of information science to evaluate the rationale. One respondent in agreeing said, "This is both a computer science and information science discipline problem." Another said: "Vocabulary control is within information science domain." A third respondent said yes, but: "also other sciences, e.g., computer science." The respondent who circled both answers remarked: "Information science is *one* of the parties in this determination." Two respondents who disagreed made comments: One declared, "Again, 'best attained' is a value judgment, not a research question for information science." The other stated: "Although information science is involved, the major expertise is in computer science." One observation is valid here. In spite of the answer given, those making comments indicated ambivalence.

Comments: As in other recommendations, respondents cited the interplay of information science with other disciplines.

E9. Several libraries, documentation centers, and abstracting and indexing services should be supported by the NSF Office of Science Information Service in efforts to develop agreed-upon canonical forms for each widely used bibliographic documentary information element.

Analysis of Recommendation

Actor(s):   NSF Office of Science Information Service and information services organizations.

Goal: Development of canonical forms for "bibliographic documentary information elements."

Rationale: Standards are needed for "bibliographic documentary information elements." NSF should support this goal. (It is not clear what these bibliographic documentary information elements are.)

Source of Rationale: Report.

Respondents—Survey Proper

	Agree	Disagree	No response
Adequacy of content analysis:	14(82.4%)	3(17.6%)	3
Information science can evaluate			
rationale's validity:	13(76.5%)	4(23.5%)	3

Survey Analysis

Adequacy of Content Analysis: There was a consensus on this question. Three disagreed, without comment.

Ability of Information Science to Evaluate the Rationale's Validity:
There was consensus regarding the ability of information science to evaluate the rationale. One who agreed said: "Partial agreement. Other actors that need to be involved are the American National Standards Institute (ANSI) and computer science. Information science does not have complete expertise." Another commented: "Bibliographic control is our middle name." The respondent who circled both answers said: "Part of this—identifying bibliographic information elements—is relevant to information science, but the value judgments are not." One who disagreed questioned: "Since the effort is supposed to be undertaken jointly, how could but one element in the series be able to effectuate changes?"

Comments: Two general observations are in order here. First, some see the possibility of applying information science along with other disciplines. Second, one respondent wanted to exclude value judgments for all applications.

CHAPTER 8

SUMMARY OF ANALYSES
OF SATCOM STUDY

From the outset this case study has had three purposes. The interrelation of these three purposes has often tended to make this simple, preliminary study into a complex one. First, and most concretely, this is a case study of the SATCOM report. In this area the study's purpose was to examine closely and identify clearly, with the aid of content analysis, important characteristics of each of the report's recommendations.

The second purpose, which is broader and less concrete, was to judge whether the content analysis applied to the SATCOM report would be applicable to other information policies. This judgment was to be based on the outcome of the investigator's analysis and its corroboration by the experts surveyed. In distinction from the first and third purposes, this one is concerned solely with the method employed in this study.

The third purpose of the study, which proved to be more abstract and more elusive than the other two, was to gain some understanding of the applicability of information science to the evaluation of information policy. Some body (or bodies) of knowledge lies at the heart of information policy evaluation, and information science was part of this body of knowledge that could be applied to carry out such evaluations. In order to determine whether this supposition had any merit, experienced practitioners in the field of information science were surveyed in order to see whether information science principles could be applied in evaluating the validity of the reasons for proposed SATCOM policy tasks.

In order to determine that these two types of analysis—content analysis of each policy task and judgment about the ability of information science to evaluate each policy task rationale—were valid, a survey (with pretest) was carried out. This survey was designed to determine: (1) if the content analysis was an adequate tool to apply to SATCOM recommendations in order to specify common elements among recommendations, and (2) if there was consensus among practitioners of information science about the ability of that discipline to evaluate a SATCOM policy task rationale. The responses and comments from the respondents to this survey gave the investigator both the ability and the authority to discuss findings regarding content analysis of SATCOM (purpose 1), the role of information science in SATCOM policy analysis (purpose 3), and the applicability and relevance of the methods used to analyze other information policies (purpose 2). This chapter reports the aggregate findings of these various analyses. The findings, however, are more suggestions of possibilities than statements of empirical truth.

This chapter is divided into four parts. Part 1 describes the results of the content analysis: The sources of rationales, actors, and goals for *all* the policy tasks are categorized and counted.[1] The incidences of actors and goals are compared to the incidences of rationale source types in order to discern if any meaningful patterns exist. This part will serve to fulfill the first purpose of the study, an examination of the SATCOM Report. Part 2 describes the responses to the survey regarding the adequacy of content analysis. It serves as a partial validation of the content analysis carried out in Part 1 and serves to fulfill the second purpose of this study, which is to determine the applicability of the method to other information policies. Part 3 describes the responses to the survey regarding the ability of information science to evaluate policy task rationales and serves to fulfill the third purpose of the study, which deals with the applicability of information science in evaluating information policy. In addition, the incidences of consensual and nonconsensual (i.e., consensus or nonconsensus regarding the ability of information science to evaluate the rationale) rationales are compared with the incidence of actor types, goal types, source of rationales, and adequacy of content analysis, all identified earlier. Part 4 discusses the limitations and constraints of the study.

PART ONE: SUMMARY OF SATCOM CONTENT ANALYSIS

Content Analysis—General Observations

Content analysis was carried out on the 55 recommendations of the SATCOM report in order to achieve the first purpose mentioned above. The goal of the content analysis was to identify important characteristics about

---

[1]"All" refers to all 55 recommendations of the SATCOM report, not just those 15 included in the survey. For an analysis of the remaining 40 recommendations, see Burger (1988).

each recommendation. These characteristics were: actor(s) who were to carry out the recommendation, the goal(s) of the recommendation, the rationale for each actor—goal combination identified, and the source(s) for the rationale. Each actor-goal combination formed what was referred to earlier as a policy task. Within the 55 recommendations, 57 policy tasks were identified, each with its own rationale. Other than this aggregate figure of the number of policy tasks, the only other aggregate statistic of interest here is the frequency distribution of the various *sources* of each policy task's rationale. This is important because it provides a way of determining how explicit SATCOM was about the reasons for its recommendations. It will also be used in discussing other aspects of content analysis. There were six potential sources or source combinations: the recommendation itself, the recommendation and the report, the recommendation and inference, the report, the report and inference, and inference alone. Table 8–1 presents the sources of rationales and each type's frequency.

The rationale was completely inferred in 5 (9%) cases, and partially in 14 (25%) cases. Some 91% of the rationales for recommendations were stated either in the recommendation itself or in the accompanying report, but only 66% of the rationales were explicit enough so that the reader (an experienced practitioner) did not have to infer any part of a rationale. Furthermore, in only 19% of the cases was the rationale stated in the recommendation itself. This suggests that rationales may usually be found in accompanying documentation or background papers,which in turn suggests that study and analysis of such policies will probably involve various methods of qualitative research techniques or combinations of qualitative and quantitative techniques in order to determine whether the disciplinary foundations for a policy recommendation are sound.

Other Elements of Content Analysis

The following subsections will summarize the results of the content analysis of all 57 policy tasks. By way of illustration, the actual content analysis of 15

Table 8–1. Source of Rationales and Their Frequency Distribution

Source	Frequency
Recommendation itself	4
Recommendation and report	4
Recommendation and inference	3
Report	30
Report and inference	11
Inference alone	5
Total:	57

Table 8-2. Frequency Distribution of Type of Actor

Type	Frequency # %
Professional societies	20 (29.85)
Anybody who deals with STINFO	8 (11.9)
Federal agencies	20 (29.85)
Federal coordinating bodies/commissions	7 (10.4)
Publishers of scientific publications	4 (6)
Unspecified actors	3 (4.5)
Researchers	2 (3)
Indexing and abstracting services	2 (3)
Employers of scientists and technologists	1 (1.5)
Total:	67 (100)

of these recommendations are found in in Chapter 7 (for recommendations A2, B5, B10, B14, C4, C6, C7, C10, C11, D2, E2, E5, E7, E8, and E9).[2]

Types of Actors

An actor, as identified in Chapter 6, is the person, institution, or group designated in the policy task to implement it. Actors are subject to lobbying, to compromise, and to other pressures that can override the persuasiveness of disciplinary knowledge. The categorization of actors can identify the various stakeholders in the policy recommendation. On the basis of the content analysis already carried out, nine types of actors were identified and their frequencies tabulated. Several recommendations specified more than one type of actor, and, therefore, recommendations could be counted more than once in this frequency distribution, provided in Table 8-2.

Some of these categories are not mutually exclusive, and this lack of distinctiveness in labeling is due, in part, to the language used in the recommendations and the report. For example, the "Anybody who deals with STINFO" category could conceivably include any of the other categories. It does not. In only includes actors who could not be classified into the other categories.

The categories and their frequencies are particularly informative, however. For example, 27 (40.25%) of the actors were associated directly with the Federal government either as employees of executive branch agencies or as appointed members of commissions and coordinating bodies. Federal coordinating bodies and commissions are generally ad hoc bodies set up by Federal agencies or Congress to carry out specific missions. Their existence

---

[2]Content analysis of the remaining recommendations can be found in Burger (1988).

may last for a few months and then disband (such as the Tower Commission) or continue indefinitely (such as the National Commission on Library and Information Science).

In the case of SATCOM the high incidence of actors belonging to professional societies or the Federal government also implies that the policy makers assumed most of the STINFO policy action would be carried out by the Federal government or professional societies acting in concert with their own professed aims, as well as for the commonweal. This type of statistic would have relevance for comparative studies of STINFO policies. Other types of information policies may assume more action taken by private enterprises, individuals, or other groups. Such a comparison on this basis with other information policies has not been carried out.

This relatively high incidence also demonstrates the internal consistency of SATCOM. Its primary recommendation, which was never carried out, was the creation of a joint commission on scientific and technical communication. The joint commission's purpose would have been to coordinate activities both within and between the public and the private sectors. It is entirely appropriate, therefore, that such a high incidence (70.1%) of the named actors are clearly labeled as Federal bodies of one sort or another and as professional societies. This was clearly a top-down plan, aimed at the political leadership and bureaucratic organizations other than at those directly involved in STINFO production or dissemination.

Actor Types and Rationale Source

Since the incidence of fully or partially inferred rationales would be significant for achieving the first purpose of this study, the investigator sought to determine if any meaningful patterns existed in the relationship of actor types and fully or partially inferred rationales. Table 8–3 presents the coincidence of type of actor with the number of rationales that were fully or partially inferred.

The incidence of partially or fully inferred rationales is roughly in proportion to the incidence of types of actor (see Table 8–2), the highest incidence occurring with Federal government and professional society actions. These categories account for 70% of the total recommendations, so the high numbers of inferred rationales (16 or 80%) are proportional to the incidence of actions in these categories. One may infer from this that SATCOM felt less compelled to provide a rationale for policy tasks to be carried out by federally affiliated bodies and professional societies: The entire thrust of the report was toward coordination of STINFO related actions using the existing power and prestige of these actors. This grand rationale pervades the report and, therefore, SATCOM may have perceived its reiteration in each specific case to be redundant.

Table 8-3. Relationship of Actor Types and Fully or Partially Inferred Rationales

Type of Actor	Incidence of Inferred Rationale	
	Partial	Full
Professional societies	6	1
Anybody who deals with STINFO	1	0
Federal agencies	5	1
Federal coordinating bodies/commissions	1	2
Publishers of scientific publications	1	0
Unspecified actors	0	0
Researchers	1	0
Indexing and abstracting services	0	1
Employers of scientists and technologists	0	0
Total:	15	5

The four inferred rationales that do not fall into these categories, however, can be explained by similar reasoning. In three of the four cases (Indexing & abstracting services, Researchers, and Publishers) the functions of each actor are fairly narrow and specific compared to the higher incidence categories. In these areas rationales for actions may have been omitted because rationales for their action may have been assumed to reside in their clearly labeled function. As with the previous table, these statistics and their related inferences may have more relevance to comparative studies.

Types of Goals

A goal, as identified earlier in Chapter 6, is any statement that presents the purpose or objective of the policy task. The identification of goals is valuable since their categorization can aid in articulating the major thrusts of the entire policy. In addition, if the goals are tightly packed into similar categories, it may indicate a resolve on the part of the policy makers to achieve a well-defined end. On the other hand, goals widely dispersed over several categories might indicate a lack of focus on the part of the policy makers, or a public policy problem so large and multifaceted that one policy may not be successful in solving it.

In a similar manner, and with a similar purpose, the investigator classified the 57 goal statements into 16 categories. Unlike the analysis of actors, where a recommendation could specify more than one type of actor, the number of goals explicitly stated is equal to the number of policy tasks identified through content analysis. Table 8-4 presents these 16 types of goals with their frequency.

In this table the 16 goal types have been arranged in six smaller groups. The first and largest group deals with ensuring the efficient processing of information products. This is comprised of evaluative and cost studies, effects of

Table 8-4. Goal Types and Their Frequency

Goal Type	Incidence # %
Ensure efficient preparation of information products/ carry out evaluation/perform cost studies	21 (36.84)
Study information economics	1 (1.75)
Report on effect of technology	1 (1.75)
Provide information services	7 (12.28)
Provide dissemination of information	6 (10.53)
Improve informal communication	3 (5.26)
Develop an information system	1 (1.75)
Provide financial support	7 (12.28)
Improve management of information programs	2 (3.51)
Develop marketing techniques	1 (1.75)
Keep abreast of standards	1 (1.75)
Share responsibility for information program	1 (1.75)
Establish organization	1 (1.75)
Develop international cooperative programs	2 (3.51)
Have representation on delegations to foreign nations	1 (1.75)
Support library education	1 (1.75)
Total	57 (100)

technology, and the study of information economics. This is, by far, the largest category and accounts for 23 or 40.35% of the total goal types. This high incidence of evaluative study and process monitoring seems to counter, to some degree, the main declared goal of SATCOM and subsequent STINFO reports—that a centralized coordinating body is needed.

An analyst might reason that in order to demand coordination as SATCOM did, the means by which to coordinate would be known. But why then would the committee turn around and demand all these evaluative studies of the process—a demand that implies less than adequate knowledge of that which is to be coordinated? In looking more closely at the coordination demand, however, an analyst would realize that this was a vague notion at best that must have sounded respectable in policy formation circles, but fuzzy among the implementers. The high incidence of these process oriented policy goals shows that SATCOM also knew that these types of studies would be needed prior to any coordination. The high incidence of these types of policy goals also implies that relatively little was known about scientific communication. If more were known, there probably would not have been such a great need to carry out evaluative studies. Exploratory and evaluative studies are usually done when a clear course of action does not present itself. Such studies may also be used to deflect immediate attention away from an issue, pending the results of the study. The former condition seems to exist here. An exploration of these possibilities would be useful in future studies.

Another large group of goals deals with the provision of information

services or improvement of scientific communication with direct suggestions. This group accounts for 17 (29.8%) of all the goals. The third largest
group (7 or 12.3%) consists of one category, the provision of financial
support, which is related to the next category. The fourth group is concerned
with management issues in scientific communication (6 or 10.5%). Given the
SATCOM concern with coordination, this seems to have been neglected in
individual recommendations. The fifth group is related to international goals
(3 or 5.26%). Within this category it is worth noting that neither of these
goals deals with the balance now sought between scientific communication
and national security. In fact, the two recommendations in this fifth group
seem to run counter to these present-day concerns about secrecy. SATCOM
saw the value and apparently minimized the danger of promoting the international flow of STINFO. This concern with secrecy is not a new one but
apparently was not a major factor in STINFO policy at the end of the 1960s,
otherwise it probably would have emerged in this survey. This suggests that
both ideology and societal values that prevail at the time of a policy proposal
may have a direct influence on the formulation of that public policy. This also
requires further investigation. The last group contains only one goal that
supports a broad base of library education.

Goal Types and Rationale Source

The investigator also sought to determine whether any pattern emerged with
regard to the type of goal and source of rationale. Table 8–5 presents the
coincidence of the goal type and of partially or fully inferred rationales.

The relative distribution of inferred or partially inferred rationales is
comparable to the frequency distribution of goal types. Given our meager
knowledge about the relationship between stated goals and rationales provided, the reason these two distributions resemble each other so closely may
be that when a proposed goal is part of a group of goals that predominate in
a policy, there is a greater likelihood that a rationale will not be provided.
There is one possible anomaly in this group of goal statements, however. The
single goal statement "Provide information services" accounts for 7 (12.3%)
of the goal types, yet has 4 (21.1%) of the inferred rationales. This may
indicate a clear idea about such a goal on the part of the policy formulators.
Just as with predominating goals, if the rationale for a goal is common
knowledge or an assumption, it will probably not be stated. What does "the
provision of an information service" mean?

PART TWO: VALIDITY OF INVESTIGATOR'S ANALYSES

In order to determine whether the investigator's content analysis possessed
validity, a corroborative survey was carried out among people who were

Table 8-5. Coincidence of Goal Type and of Partially or Fully Inferred Rationales

Goal Type	Partially #	Fully #
Ensure efficient preparation of information products/carry out evaluation/perform cost studies	6	1
Study information economics		
Report on effect of technology	1	
Provide information services	2	2
Provide dissemination of information	1	
Improve informal communication		
Develop an information system		
Provide financial support	2	1
Improve management of information programs		1
Develop marketing techniques		
Keep abreast of standards		
Share responsibility for information program	1	
Establish organization		
Develop international cooperative programs		
Have representation on delegations to foreign nations		
Support library education	1	
Total:	14	5

considered experts in the field of information science. This part of the analysis is directly concerned with this study's second purpose dealing with method.

After examining responses to the pretest, and having been satisfied by those results that the method was sound, the investigator then randomly (using a random number table) chose 15 different SATCOM recommendations and sent these out to 45 different people, with the same two questions asked for each policy task: (1) Was the content analysis, which identified the actor(s), goals, rationales and source, adequate?; and (2) Could information science, as the respondent understood it, be used to evaluate the validity of each rationale presented? This group consisted of 35 males and 10 females. Some 21 of the members, 19 men and 2 women, were either on the original SATCOM committee or on its board of consulting correspondents.

A total of 20 respondents (44%) returned questionnaires, and 5 (10%) others replied with letters apologizing for not responding to discussions of the survey itself. The individual responses to the questionnaire are reported in Chapter 7. The content of the five letters will be summarized below, after the reporting of the survey results.

Adequacy of Content Analysis: Survey Responses

Table 8-6 reports the summary responses to the question concerning the adequacy of the content analysis and is related to the first and second purposes of the study.

Table 8-6. Adequacy of Content Analysis, Survey Responses

Recommendation	Adequate # (%)	Inadequate # (%)	No Response
A2 (improve management of STINFO)	19 (95)	1 (5)	0
B5 (make public information available to applied scientists)	15 (75)	5 (25)	0
B10 (evaluate and improve abstracting services)	18 (90)	2 (10)	0
B14 (identify IACs and make their services more widely available)	18 (90)	2 (10)	0
C4 (develop guidelines for author-supplied documentation units)	20 (100)	0 (0)	0
C6 (support program in library education)	16 (84.2)	3 (15.8)	1
C7 (analyze economic aspects of formal sci-tech publications)	16 (84.2)	3 (15.8)	1
C10 (experiment with different ways of publishing research results)	17 (89.5)	2 (10.5)	1
C11 (publish meta information)	17 (94.5)	1 (5.6)	2
D2 (provide opportunities for interinstitutional visits of scientists)	16 (88.9)	2 (11.1)	2
E2 (develop measures of value for info services)	15 (83.3)	3 (16.7)	2
E5 (effectively combine machine and human functions in info retrieval)	13 (72.2)	5 (27.8)	2
E7 (1st goal) (experiment with technology in communication of STINFO)	17 (94.4%)	1 (5.6%)	2
E7 (2nd goal) (employ experts in experimenting with technology in communication of STINFO)	16 (88.9)	2 (11.1)	2
E8 (develop and evaluate file-format languages)	15 (83.3)	3 (16.7)	2
E9 (develop canonical forms for "bibliographic documentary information elements")	14 (82.4)	3 (17.6)	3
Total	262 (87.33)	38 (12.67)	

The responses ranged from 100% to 72.2% agreement that the content analysis was adequate.

The average percentage of agreement, 87.33%, indicates a relatively high degree of satisfaction with the content analysis relative to the arbitrarily chosen scale of 70% discussed earlier. From the comments given to individual responses (see, for example, the responses to recommendations C10, E2, E5 in Chapter 7), some of the measured dissatisfaction was, in fact, directed at the recommendation itself and not at its content analysis. (C10 deals with quicker and more efficient means of publishing research results; E2 suggests the development of value measures for information services; and E5 urges continued experimentation in the design and use of effective combinations of machine and human factors in information retrieval.) Still, it does indicate that while the application of content analysis is adequate, it could still improve.

Applicability to Other Information Policies

From the measured *satisfaction* with the content analysis, as well as the ability to catalog the various actors and goals, it is reasonable to suggest that any information policy would benefit from this type of content analysis. Any tool that makes a policy more understandable is beneficial for the study and improvement of that policy.

Difficulty in Determining Rationale

As comments to specific recommendations implied, there was some question concerning the proper identification of the rationale for these recommendations, especially in those cases where the rationale was inferred wholly or in part. This difficulty arises from two sources. First was the difficulty of identifying the proximate rationale for each task: In several cases it was difficult to distinguish proximate and other rationales. Second, the rationale as stated may not have been worded clearly, thereby giving rise to doubt whether information science was able to evaluate it.

PART THREE: ABILITY OF INFORMATION SCIENCE TO EVALUATE RATIONALES

The survey also questioned whether the discipline of information science encompassed the knowledge necessary to evaluate the rationale for each policy task. The answers were open to interpretation. The main difficulty here is that information science does not have clearly defined boundaries. Hence, some respondents might differ regarding their notion of what the discipline of information science is, either by having a very confused notion of the discipline (see summaries of letters below) on the one hand, or by encompassing (for example) all knowledge within the discipline of information science, on the other hand. Therefore, the following survey responses should be examined with these caveats in mind. According to the guideline established earlier, consensus is indicated when 70% or more of the respondents give the same responses (i.e., 70% or more agree or 70% or more disagree).

Tables 8–7A and 8–7B present the results of the second question of the survey, regarding the ability of information science to evaluate the rationale for each policy task.

Table 8-7A. Ability of Information Science to Evaluate Rationales
(Recommendations Having a Consensus)

Recommendation	Can Eval # %	Cannot Eval # %	No Response
B5	14 (70)	6 (30)	0
C4	14 (70)	6 (30)	0
C6	15 (78.9)	4 (21.1)	1
C10	14 (73.7)	5 (26.3)	1
C11	15 (78.9)	4 (21.1)	1
E2	17 (89.5)	2 (10.5)	1
E5	18 (100)	0 (0)	2
E8	13 (72.2)	5 (27.8)	2
E9	13 (76.5)	4 (23.5)	3
Total:	133 (78.7)	36 (21.3)	16

Table 8-7B. Ability of Information Science to Evaluate Rationales
(Recommendations Not Having a Consensus)

Recommendation	Can Eval # %	Cannot Eval # %	No Response
A2	11 (55)	9 (45)	0
B10	13 (65)	7 (35)	0
B14	11 (55)	9 (45)	0
C7	11 (55)	9 (45)	0
D2	7 (38.9)	11 (61.1)	2
E7 (1st goal)	13 (68.4)	6 (31.6)	1
E7 (2nd goal)	13 (68.4)	6 (31.6)	1
Total:	79 (58.1)	57 (41.9)	

Respondents reached consensus on 9 of the 16 policy tasks that they were asked to judge. Consensus ranged from 100% agreement (Recommendation E5, dealing with the design and use of effective combinations of machine and human functions in information retrieval) to 70% (Recommendation B5, urging the creation of information programs that will make public documentation easily available to practitioners of applied science, and C4 dealing with the development and promulgation of guidelines for author supplied documentation units). The average response was 78.7%. Only two responses were over 79%: E5 (see above) and E2 (development of measures of value for information services). In addition, these two recommendations were within group E of the SATCOM recommendations, which dealt in general with Studies, Research, and Experiments—an area where one might expect information science to have the most applicability. Compare, for example, group A (Planning, Coordination, and Leadership at the National Level), where a policy analyst might not expect information science to be applicable at all since most of the recommendations are political in nature. But no one should interpret SATCOM groupings as a sign demonstrating the applicability of information science for evaluating rationales. The categories used in SATCOM were not formulated for this purpose, but simply for grouping a set of recommendations for action. Furthermore, in all cases where consensus was reached, the consensus was that information science *could* evaluate the rationale. Only in one case, D2, dealing with interinstitutional visits of scientists and technologists, was the *tendency* to agree that information science could not evaluate the policy task apparent. This may indicate an overblown sense of the range of the discipline of information science, or may be due to the character of the policy tasks sampled. Further research is necessary here in order to make a confident judgment.

Respondents did not reach consensus on 7 of the 16 policy tasks they were asked to judge. Three of the responses were almost split evenly: A2 (dealing with management of STINFO), B14 (identification and development of information analysis centers), and C7 (analysis of economic aspects of formal STINFO publications). Two were very close to the cutoff point: E7 (1st and 2nd goals) (use of advanced technology in scientific-technical communication), and D2 (interinstitutional visits of scientists and technologists) was the only one to exhibit a tendency of consensus away from the applicability of information science.

In general, all but the responses to E2 (89.5%) and E5 (100%) exhibit a lot of disagreement concerning the intellectual strength of the discipline of information science. Some of this disagreement may be attributable to the ambiguous language of the recommendations, but some may also be traced to the as yet undetermined nature of information science itself.

SUMMARY OF CONTENT OF RESPONDENTS' LETTERS

Five additional respondents sent letters explaining the reasons for their non-response. One was a courtesy letter that cited lack of time for the respondent's inability to respond. Two other respondents declined because they perceived information science as a technical field, and were unable, therefore, to judge these public policy questions.

An academic respondent refused to reply on the principle that he did not "participate in perception surveys that use the pretentious phrase 'information science' when respondents are given free rein to exercise their wildest wishes and aspirations."

A person in an information business declined for two reasons: 1) "most of the subjects under discussion are far out of context and might be true in the context of 1969 and untrue and insignificant in 1987"; and 2) "many of the rationales have long since been analyzed and acted upon, very frequently by

means or routes other than those suggested in SATCOM and your study."
He cited as an example recommendation B10, dealing with indexing and
abstracting services. SATCOM, in this recommendation, urged research
sponsors to contribute to the costs of indexing and abstracting. Today, the
respondent noted, these services are his business.

There were two main points to the final respondent's letter. "Much has
changed since the SATCOM report was prepared some 20 years ago. Thus if
we are supposed to validate your analysis of the Report it would little trouble
me in circling 'agree' in almost every case." As far as the second question is
concerned he states:

> How do I handle 'agree' or 'disagree' when asked to assess the role of the 'dis-
> cipline of information science' when I do not believe that such a discipline exists.
> If there is such a discipline, the nature of its efforts are such that no meaningful
> contribution could be made in the areas you discuss. The problems of the field
> of STINFO, in fact all information, are better solved by the efforts of those
> disciplines that deal with human behavior. Thus I would circle 'disagree' on any
> contribution that could be made by the discipline of information science. But in
> order to confuse the issue more completely, I would circle 'agree' to the state-
> ment that many persons engaged in what they call information science have the
> skills and experience to make useful contributions toward the evaluation of
> policy as outlined by you.

Virtue of Explicit Rationales

If the reasons for a policy decision are not made clear in the policy document
itself, then the discernment of the validity of the recommended policy action
becomes difficult and often speculative. This has implications not only for
policy formation but also for policy implementation and post-policy evaluation. As far as policy formation is concerned, if neither critics nor supporters
of a policy know the rationale upon which a recommended action is based,
then it will be difficult to judge the future effectiveness of that recommendation. For policy implementation, it has been shown how bureaucracies have
molded policies to their own ends because, in some cases, the intent of a policy
was not explicit, nor the goals made clear. For post-policy evaluation, if rationales are not evident or stated, then the intent of the policy will also not be
evident. Measures to correct the failures of the policy will be based solely on
what the policy analysts thought the policy should have intended to do. Of
course, policy makers will intentionally obscure the reasons for policy actions
by providing bogus rationales (as in the Iran–Contra affair), but such rationales can at least be evaluated to determine their validity. Their one virtue is
that they are explicitly stated.

Further Analysis

In an attempt to find out whether consensual agreement about the ability of
information science to evaluate a policy task rationale bore any relationship to

other factors identified in the study, an examination of the coincidence of consensual or nonconsensual recommendations with types of actors, types of goals, source of rationale, and adequacy of content analysis was carried out. Table 8-8 presents the coincidence of consensual and nonconsensual recommendations with type of actor.

The frequency of both consensual and nonconsensual recommendations is proportional to the gross number of recommendations classified under each type. Since the number of cases is not large enough to make any statistically significant assertions, the following observations are suggestive only. There do seem to be some differences among types of actors and the potential for consensus or nonconsensus. For example, one might expect that the rationales given for actions performed by indexing and abstracting services or researchers could be evaluated by information science. The survey responses support this expectation. Furthermore, another expectation, also supported, is that actions carried out by Federal coordinating bodies or commissions would be primarily political in nature. Finally, those four categories with both consensual and nonconsensual responses might be looked at in several ways depending on the clarity and nature of the recommendation itself.

But whether these speculations would be supported if all the recommendations had been examined by experts is another matter. The rationale is based on the *policy task*, which consists of an actor and a goal. To divorce actor and goal makes the task relatively meaningless in such a coincidence table because it removes a critical part upon which the rationale and its evaluation are based. The same problem occurs in Table 8-9.

Some goals dealing with study, management, and financial support lacked consensus. This suggests that for SATCOM rationales for some goals for these areas lie outside the evaluative boundaries of information science. But this suggestion is subject to at least two caveats. First, as with Table 8-8 the focus is only on part of a policy task. To judge the whole by a part can be misleading. Second, the language used to describe the goal might sound more

**Table 8-8. Actor Type and Coincidence of Consensual (C)
and Nonconsensual (NC) Recommendations**

Type	Freq	C	NC
Professional societies	20	3	1
Anybody who deals with STINFO	8	1	1
Federal agencies	20	3	1
Federal coordinating bodies/commissions	7	1	
Publishers of scientific publications	4		
Unspecified actors	3	1	1
Researchers	2	1	
Indexing and abstracting services	2	1	
Employers of scientists and technologists	1		1
Total:	67	10	6

Table 8-9. Goal Type and Coincidence of Consensual (C)
and Nonconsensual (NC) Recommendations

Goal type	Freq	C	NC
Ensure efficient preparation of information products/ carry out evaluation/perform cost studies	21	6	2
Study information economics	1	1	
Report on effect of technology	1		
Provide information services	7		1
Provide dissemination of information	6	2	
Improve information communication	3		1
Develop an information system	1		
Provide financial support	7		1
Improve management of information programs	2		1
Develop marketing techniques	1		
Keep abreast of standards	1		
Share responsibility for information program	1		
Establish organization	1		
Develop international cooperative programs	2		
Have representation on delegations to foreign nations	1		
Support library education	1	1	
Total:	57	10	6

amenable to evaluation by information science if certain keywords or buzz-words are used. An interesting experiment would be to produce two texts describing the same policy task. One would be loaded with buzzwords often used in information science research. The other would not use such words. How would our experts judge these?

Two expected occurrences of consensus do appear with the provision of the dissemination of information and ensuring the efficient preparation of information products/evaluation. In spite of these suggested conclusions, however, the table seems to raise more questions that it answers.

The results of Table 8-10 show no pattern. All but one of the 16 rationales have split responses. Apparently the source of rationale bears no relationship to the incidence of consensus regarding the ability of information science to evaluate the rationale.

Table 8-10. Coincidence of Source of Rationale and Consensual (C)
and Nonconsensual (NC) Recommendations

Source	Freq	C	NC
Recommendation itself	4	2	1
Recommendation and report	4		1
Recommendation and inference	3		
Report	30	4	2
Report and inference	11	3	3
Inference alone	5		
Total:	57	9	7

Table 8-11A. Coincidence of Adequacy of Content Analysis and Consensus of Ability of Information Science to Evaluate the Rationale (Recommendations Having a Consensus)

Recommendation	Adequate # %	Inadequate # %
B5	15 (75)	5 (25)
C4	20 (100)	0 (0)
C6	16 (84.2)	3 (15.8)
C10	17 (89.5)	2 (10.5)
C11	17 (94.4)	1 (5.6)
E2	15 (83.3)	3 (16.7)
E5	13 (72.2)	5 (27.8)
E8	15 (83.3)	3 (16.7)
E9	14 (82.4)	3 (17.6)
Total:	142 (85.03)	25 (14.97)

For consensual recommendations (see Table 8-11A), the adequacy of content analysis response ranged from a high of 100% agreement about its adequacy (C4, guidelines for author-supplied documentation units) to a low of 72.2% (E5, experimentation in the design and use of effective combinations of machine and human functions). These were the extremes for both scales. The average for consensual recommendations was 85.03%.

For nonconsensual recommendations (see Table 8-11B), responses had a smaller range, from a high of 95% (A2, dealing with improved management of STINFO programs) to 84.2% (C7, systematic analysis and study of the economic aspects of formal STINFO publications). The average here (90.25%) was higher than that of consensual recommendations.

It is anomalous that in areas where a consensus was reached regarding the applicability of information science for evaluation that adequacy of the content analysis was judged to be somewhat lower. This may be due to the

Table 8-11B. Coincidence of Adequacy of Content Analysis and Consensus of Ability of Information Science to Evaluate the Rationale (Recommendations not Having a Consensus)

Recommendation	Adequate # %	Inadequate # %
A2	19 (95)	1 (5)
B10	18 (90)	2 (10)
B14	18 (90)	2 (10)
C7	16 (84.2)	3 (15.8)
D2	16 (88.9)	2 (11.1)
E7 (1st goal)	17 (94.4)	1 (5.6)
E7 (2nd goal)	16 (88.9)	2 (11.1)
Total:	120 (90.23)	13 (9.77)

fact that if a recommendation is analyzed well, specialists will more readily be able to know when information science is applicable. On the other hand, when a recommendation is analyzed poorly, there seems to be a tendency to assume the applicability of information science. This phenomenon was manifest also in the tendency of the experts to have the direction of consensus (in all but one case) towards the applicability of information science. This may be indicative of an overblown sense of information science's power and applicability.

PART FOUR: LIMITATIONS, CONSTRAINTS, AND ASSESSMENT OF CASE STUDY

General

Contemporaneity.   One limitation that inhibited this analysis and which would probably hinder any policy analysis,to some degree, was that the investigator was only privy to the final published document called the SATCOM report. If the issuance of the report were a contemporary event, the analyst might also have been able to clarify certain points of interpretation or learned unreported reasons for specific policy tasks.

Language.   Closely related to this last point is the language problem. The language of the recommendations and the report is often obscure and jargon-ridden, and because of this there is difficulty sometimes in determining exactly what a recommended policy task is supposed to accomplish. For example, the proposed actor for a policy task could be obscure or unspecified (e.g., recommendations A1, B8, C7, E1-E3), the entire recommendation could be generally obscure (e.g., A3), or the intended goal or actors could be very broadly cast (e.g., B15, B16, C1, C3, D2). If any part of the proposed policy task is unclear, then it will be difficult to evaluate the worth of that task.

The Discipline of Information Science.   Using "information science" was risky and did lead some of those surveyed to refuse to answer the questionnaire. From the responses to the survey of experienced practitioners, it is evident that not all agree on the nature of information science, nor do they agree on the applicability of information science to specific recommendations of the SATCOM report. But one does not have to take the radical position of Schrader (1984) that no theoretical progress can be made in information science until the terminological confusion is erased and consensus is reached on the nature of information science. The progress of knowledge has not always proceeded in this way. Indeed, although the mental models of those surveyed are obviously not coincidental, it by no means obscures the fact that information policy cannot be evaluated by information science, whatever it is, alone. But it does play a yet undefined role.

Nature of the Public Policy Problem.   Another factor that presented difficulties was the nature of problems to be solved by public policy. By and large, problems addressed by public information policy are problems on a macro scale, involving aggregates of systems. Information science, on the other hand, essentially deals with micro problems, problems on a system level within one organization. Other disciplines, such as sociology and economics, have developed to include the macro and micro scales. There is little or no evidence that information science has done the same. Until it does, if it does, its relevance to the analysis of information policy seems to be limited.

Values Inherent in Analysis.   Finally, one limitation not discussed in the first chapter of this study is the presupposition that analysis of the SATCOM report, or any information policy for that matter, can be purely objective. Such analyses also involve values and ideology. This is an area that needs further investigation.

On the basis of the analyses of individual recommendations, the comments made by those surveyed, and the results of the foregoing data summary, several conclusions have emerged from this case study with respect to the SATCOM report—in particular, the method used; information policy in general; and the nature of the role that information science could play in the evaluation and design of information policy.

SATCOM Report

As far as the SATCOM report itself is concerned, this analysis has confirmed assessments of the report contemporary with it (see Chapter 6). The SATCOM report possessed at least two serious flaws that may be endemic to information policy in general. The two flaws are the vagueness of goals and the assumption of the applied science analytical paradigm in dealing with information policy problems and their solutions.

Other defects, however, may also be endemic to information policy making in general. One of these is the ambiguity of the language used in the SATCOM report. Ambiguous language poses hermeneutical problems concerning the interpretation of each of SATCOM's recommendations. With unclear and imprecise language, implementation and evaluation are extremely difficult, if not impossible.

Another flaw, uncovered by the systematic application of content analysis, was the reliance on a political mechanism, a coordinating body such as the proposed Joint Commission, that was not described more thoughtfully.

Aside from generalizations about the need for coordination, there was never any specific mention about what that coordination would actually entail. Moreover, SATCOM sought to give the body legitimacy through status rather than through authority. The application of political science to the evaluation of recommendations urging such steps might have avoided such pitfalls.

Finally, because of all of these flaws, it was nearly impossible to apply information science knowledge to evaluate in detail specific policy tasks or to predict their outcome. All that could be done for each recommendation was to specify what knowledge might be applicable. But since few parameters were specified, it was difficult to put that knowledge to work.

Aside from these negative generalizations about SATCOM, the study has resulted in several specific findings regarding the report's characteristics.[3] One surprising finding relates to the absence of rationales for actions taken. In several cases (14 out of 55, or 25.4%) the rationale had to be inferred in whole or in part. We might not expect rationales for policy tasks to be stated explicitly in legislation, but in a document that recommends policy actions, which relies on current research and expertise, it is surprising that the reasons for proposed actions were so frequently absent. One possible reason for this is that rationales for actions are not often found in policy proposals but rather in accompanying documentation, background papers, or, in the case of actual legislation, in the hearings held to collect information about an issue. Rationales tend to be implicit in areas emphasized by the policy, that is, in areas that the policy makers consider to be the heart of the proposal. In the case of SATCOM, for example, little explanation was given for the main thrust of the policy—the establishment of a coordinating body (the Joint Commission). But whenever coordination was discussed or its need invoked, the only rationale was an indication of coordination's absence. What coordination would actually entail, or what its absence actually signified, was never discussed in any detail. It was apparently assumed that readers of the report would know what coordination entailed. A corollary of this, also revealed by content analysis, is that there is a tendency not to provide a rationale for policy tasks when the goal of that task is a part of a group of goals that predominate in a policy—in this case goals that focused on coordination at the national level.

Rationales also tend to be implicit when an actor's role is obvious or narrowly focused. An example is the activity of abstracting and indexing services. A related finding is that the rationale will not be provided when it is

---

[3]The descriptive statistics cited in the remainder of this chapter were compiled on the basis of an analysis of the entire SATCOM report. Analyses of specific recommendations may be found in Burger (1988).

assumed that the intended goal of a policy task is taken for granted. An example of this is the goal of "provision of information services."

Content analysis also clarified characteristics about the actors of SAT-COM. Some 40.25% were associated with the Federal government and 29.85% with professional societies, for a total of approximately 70% of the actors. These two figures demonstrate both the internal consistency of SATCOM—where the main recommendation was the establishment of a Joint Commission—and the fact that it was a top-down plan aimed at the political and professional leadership.

As far as goals are concerned, 40.35% dealt with evaluative study and process monitoring, implying that less than adequate knowledge was available about scientific communication. Further, about 30% of the goals dealt with the provision of information services. Financial support accounted for approximately 12% of the goal statements. One anomaly—management goals only accounted for 10.5% of the total—suggests an earlier contention in this chapter that although the main thrust of SATCOM was coordination, which is a managerial concern, very few explicit statements or suggestions were given regarding what this actually meant. Finally, in contrast to present day concerns, only about 5% of the goal statements dealt with international issues.

Method

Conclusions regarding the method used must be divided into three groups: those regarding content analysis, decisions based on that analysis (i.e., the ability of information science to evaluate the rationale), and the corroborative survey.

Content Analysis. There was no doubt that the application of content analysis did clarify specific characteristics about each recommendation. This, in turn, helped to point out the problems associated with SATCOM that were discussed in the previous section. In examining the completed analysis, however, some other problems do emerge. First, in some cases it was difficult to determine who the actor was. This occurred in spite of the reasonably precise definition of "actor" given in Chapter 6, and because of the nature of the written report itself. For example, an analysis that is forced to infer content may not be verifiable. The inferential act itself is subjective. The question of interpretation remains. What suffers in such cases is the reliability and verifiability of the method. Second, although goals were relatively easy to identify, content analysis did not help to clarify them. Often the goal statement was lifted from the recommendation itself; in other cases it came from the report. But if the goal was vague, accurate identification did not change its vagueness. Finally, in several cases, the investigator did have trouble in identifying precisely what the proximate rationale was. In 19 out of 57 (33.3%) cases it was inferred in whole or in part. The difficulty here may be traced to several things, among them the lack of clarity in the wording of the report, a possible lack of understanding about the phenomenon that was to be controlled by the recommendation, the difficulty in sorting out proximate from nonproximate rationales, or other as yet unidentified reasons. But on the whole, given the nature of policy, content analysis was to be analytically beneficial.

Decisions Based on Analysis. In addition to the need for clarifying the various parts of each recommendation, one of the reasons for applying content analysis was to decide if information science could evaluate the rationale. To make this decision one must determine what the rationale is, then know what information science is, and finally decide if that disciplinary knowledge can evaluate the rationale identified. The difficulties in determining what the rationale was have been discussed in the previous section. The relationship between information science and information policy will be discussed below. What is germane here is the *definition* of information science. This has been a challenging problem throughout the writing of this book. Rather than relying on a distilled linguistic definition of information science, information science was described operationally. The areas covered by this definition have been identified in the literature. As the results of the survey suggest, however, not all information scientists surveyed ascribed to the same definition of information science.

The main point of this study, however, was not to arrive at a precise definition of information science. The boundaries of this discipline may well remain vague and undefined for many years to come. Instead, the point was to garner some evidence that elements of information science had a role to play in the evaluation of information policy. The survey responses affirm information science's role in information policy evaluation, based on this case study.

The Corroborative Survey. The major problem in the survey was the inability of the investigator to determine whether the respondents understood the instructions for completing the survey. There is some evidence in the respondents' comments to suggest that in answering the first question about content analysis, respondents may have been agreeing or disagreeing with the recommendation itself and not with the investigator's content analysis of it. Distortion of responses may have also occurred in the way the two questions were asked and the language used. From the data received, however, it is not possible to confirm if there was a distortion. One encouraging finding was that the average consensus of agreement with the content analysis was 87.33%. The application of content analysis could, therefore, benefit other policies because it helps to clarify the content and rationale of a policy. It would also allow information policies to be compared with one another.

Information Policy

Information policy, as represented by the SATCOM report, is often vague in its specification of explicit goals. Goals were often articulated, but were not done so in a manner that would permit evaluation of the goal if the policy had been carried out. The very fact that explicit goals cannot be stated should not deter policy makers from trying to be more specific about a policy's goals. A statement that a state of affairs should be improved does state a goal. But if there is not baseline determination of the current state, or any indication of what "improvement" actually means, how can evaluators tell whether the policy has reached its goal? Clearly, they cannot.

By vaguely defining policy goals, formulators may abdicate responsibility for the policy's success or failure to the implementers of such policies. For, when faced with vague goal statements, an implementer will naturally model the policy to his or her own goals, resources, and ideology to the extent that the implementer can reasonably show that the vague goal of the policy has been accomplished. Furthermore, vague goal statements leave policy implementation susceptible to political influences.

Moreover, what vague goals exhibit is an underdeveloped, or possibly an absence of, understanding about the total range of phenomena supposedly controlled by the information policy at hand. For example, in setting a goal such as increasing critical reviewing by qualified scientists, the policy maker may be convinced with appropriate evidence that this will increase research efficiency. What may be forgotten, or minimized, is the reward system in science that may dissuade capable researchers from writing such reviews. Without an understanding of the sociology of science, for example, and without allowance in the policy goal setting to deal directly with potential problems that may impinge on the primary goal of increased critical reviewing, the policy will be ineffective. What is needed, therefore, are either quantifiable goal statements, or, in cases where these are not possible or appropriate, clearer, objectively formulated goals, or translation of goals into objectives.

Information Policy as Cultural Policy

Many of those surveyed for corroboration of the method commented directly or indirectly on the sociological aspects of the SATCOM report. In spite of any of the technical aspects that inhere in such policies, one must conclude that information policy is a cultural policy. In spite of the amount of technology involved, information policy involves human beings, their behavior, and their values. By recognizing information policy as cultural policy, there are several implications. First, a pure science or applied science analytical paradigm is inappropriate for information policy evaluation and design. This may not be a surprising conclusion until one considers the amount of applied science methodology and thinking that has been assumed when making recommendations. The knowledge upon which many of these recommendations have been based is knowledge grounded in systems thinking and the scientific method. This mode of discourse, valid for closed microsystems such as the technical aspects of a specific information retrieval system, is not applicable for large open macrosystems, areas where information policy is designed to be effective. Systems thinking and the scientific method assume that if one part is affected in a measurable way, the effects of that stimulus can be gauged throughout the entire system. There is nothing inherently wrong with such thinking, for we humans must simplify the natural world in order to understand it. The systems concept and scientific modes of discourse, however, are often taken as a final step in many analytical processes, even when this is not appropriate. Elements that do not fit the systems model, such as irrational, counterintuitive human behavior, are often not accounted for, or relegated to a minor position in the analysis. Systems thinking often omits history, a crucial, not easily measured variable. There is little historical study of information policy either in one country or comparatively. Until there is such a study, this ignorance may be one foil in designing and evaluating good information policies.

Furthermore, different information policies can accomplish different types of goals. The way that policy formulators and implementers handle policies with different types of goals has yet to be explored in depth. Twenty years ago Lowi (1972) designed a four-part scheme to describe how Congress would deal with policies whose primary goal was either distributive, redistributive, regulatory, or self-regulatory. Only Milevski's (1986) and Chartrand's (1986a) classifications of information policy legislation go anywhere in the direction of such analysis.

Relevance of Information Science

This case study of the SATCOM report suggests that the discipline of information science, as now conceived by its practitioners and defined in this study, does play a role in the evaluation of information policy. The question that remains is, what exactly is the nature of this relationship? Certainly, it is not the only discipline useful or applicable in the design and evaluation of information policy. Indeed, respondents to the survey commented time and again about other disciplines that were better suited to the evaluation of specific SATCOM policy task rationales, or in some cases, that other disciplines, along with information science, would be appropriate.

But a caveat is in order. The model of information science used here is primarily microinformatics, that is, the principles of information theory, representations for retrieval, and models of information retrieval that were formulated within closed microsystems. Information theory has been applied to large macrosystems, but apparently with little concern about the appropriateness of its applicability there. Only bibliometric principles have proven to be applicable to open macrosystems.

Furthermore, as responses to the survey often indicated, information science seems to be defined by the applier. Different degrees of consensus or nonconsensus might have been reached had the participants engaged in an actual Delphi study where there would be opportunity for feedback and clarification of responses.

Those surveyed could only reach consensus on 9 of the 16 policy tasks. Furthermore, only 2 of these 9 were above 79% agreement. This either indicates a problem in the practitioners' mental model of information science's disciplinary boundaries, an inability to judge where existing information science knowledge can evaluate a rationale, or a combination of these and previously mentioned limitations.

At this point something should be said regarding the respondents' mental model of information science. Any attempt to perceive mental models of a discipline can only be grounded in responses to questions regarding that discipline. The answers to these questions are also probably influenced by other variables present at the time the questions were asked. Analysts clearly lack the ability to create a reliable mental model of a discipline, especially from a case study. With these caveats in mind, however, it is possible to see at least two aspects of several experts' mental models of information science that emerge here. One is the inclusion of economics as part of information science and the other is an ethical dimension to information science.

In two cases, recommendations B10 and C7, there was a lack of consensus regarding the ability of information science to evaluate the rationale. (B10 dealt with input costs for indexing and abstracting services, while C7 dealt with the study of economic aspects of information services.) This ambivalence may indicate an increasing influence of economics in the discipline, especially as it moves beyond evaluation and study of closed microsystems to large open macrosystems, such as national information policy. Economics purportedly has theories that can deal with such open macrosystems and, although the predictive value of these theories is not high, the theories do offer a semblance of theoretical and disciplinary rigor in dealing with social systems that information science apparently lacks.

Another area of ambivalence, probably also arising from the application of information science to social and cultural systems, and also tied to economic concerns, is the making of nontechnical value judgments. In closed microsystems, the values of spatial and temporal efficiency seem to reign supreme. As information science is applied to larger macrosystems, economic cost/economic benefit questions, as well as social cost/social benefit and cultural cost/cultural benefit questions, arise. In the social realm, however, the calculus of cost/benefit begs other questions of value and gives rise to ethical concerns. This type of ambivalence in the disciplinary boundaries is evidenced in recommendations C6, regarding support of library education programs, and E2, regarding measures of value for information service.

The scant evidence of ambivalent responses from one case study enables some speculation about the future development of information science. Information science is a discipline that is still in the early stages of development, and its boundaries are still fluid. How the discipline will change as its practitioners are increasingly involved with questions such as information policy evaluation and design remains to be seen. Information policy evaluation itself will also be affected by this interaction.

There is, of course, a danger in relying upon any one discipline for evaluation. Since disciplines are consensual, specialized entities, the application of information science alone, as now conceived, for example, could simply serve to pass along values assumed by information science, such as spatial and temporal efficiency. Information science, however, would not be in a position to contribute an approach to information policy that may contravene the values of spatial and temporal efficiency. Furthermore, it certainly would not supply a paradigm-critical approach to information policy, which would go beyond value choices to other ways of perceiving the problems that occasion the need for information policy. Perhaps future research will answer these questions. The final chapter offers some general suggestions for the future development of an information discipline that is relevant to the design and evaluation of information policy.

PART IV

THE FUTURE OF INFORMATION POLICY EVALUATION

CHAPTER 9

CONCLUSION

Do analysts and policy makers know what they are doing in designing and evaluating information policies? This book presents the role of specialized knowledge in the evaluation and design of information policy. Because such specialized knowledge probably did not yet exist in an intelligible, codified form, one purpose of the book was to discover some characteristics about this knowledge and specific areas of information policy where it is applicable. The book has raised questions about what that knowledge is and whether that knowledge is included in the discipline called information science. This chapter posits some tentative answers to the questions raised and suggests a broader and more inclusive role for an "information science" applicable to the problems that information policy attempts to solve. Nonetheless, the tentative answers have raised additional questions.

AN EXPANDED VIEW OF INFORMATION POLICY AND ITS EVALUATION

Chapters 1 and 2 examined the political and international contexts for information policy making. This was done because, all too often, policy analysts tend to look too narrowly at information policy and at the means for evaluating it. In fact, Pacey (1983) has maintained that this is done in a number of spheres involving technology. He suggests a broader view of any public enterprise that encompasses organization, technology and culture. His diagram of technology practice (Figure 9–1) illustrates this point.

By technology practice Pacey (1983, p. 6) means "the application of

159

Figure 9-1. Diagrammatic Definitions of "Technology" and "Technology Practice"

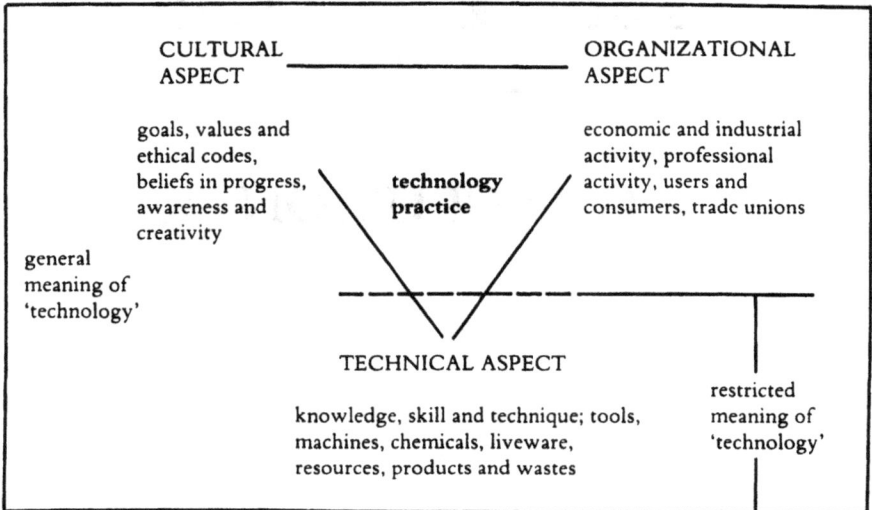

CULTURAL   ORGANIZATIONAL
ASPECT   ASPECT

goals, values and   economic and industrial
ethical codes,   activity, professional
beliefs in progress,   **technology**   activity, users and
awareness and   **practice**   consumers, trade unions
creativity

general
meaning of
'technology'

TECHNICAL ASPECT

                                             restricted
knowledge, skill and technique; tools,   meaning of
machines, chemicals, liveware,   'technology'
resources, products and wastes

Reprinted by permission of MIT Press. Copyright MIT Press, Cambridge, MA.

scientific and other knowledge to practical tasks by ordered systems that involve people and organizations, living things and machines." He clarifies this distinction between the technical aspect of technology practice and the organizational and cultural aspects by drawing an analogy with medical technology and medical practice.

Medical technology is the technical aspect of medical practice. One would not define medical practice solely in terms of the technology employed in community health care, for example. It is one important part, to be sure, but not the most important. Medical technology cannot be employed unless there are organizations in which to use it, organizational economic policies that determine what patients need to pay for services, what is a fair wage for medical service personnel, and ethical concerns that control its use in a just and humane way. Pacey maintains that if analysts and practitioners only look at one of these three aspects (culture, organization, or technology), and apply the mode of reasoning that is applicable to it to the other two, allowing it to dominate other appropriate modes of reasoning, health care policy will probably fail.

If this reasoning is extended to information policy, if only one aspect of the problem is examined in evaluating or designing it, or if only one mode of discourse, such as economics or law or scientific reasoning, is allowed to capture the imagination in dealing with information policy, then the design and evaluation will not be appropriate to the problem confronted. By "not appropriate" is meant that the solution does not put the welfare of human beings first in the design and evaluation, that the scale of the proposed solution may be too grand to achieve the desired end, and that the pace of the implemented policy is potentially harmful to human beings and human culture. This description of inappropriate is imprecise. By using such a description the intent is merely to indicate a direction, not sketch out a program.

Several writers (Toulmin, 1990; Von Laue, 1987; Ravault, 1987; Stanley, 1978) have recently shown that the predominant mode of reasoning that now exists in Western civilization is the scientific–technical mode. This mode tends to dominate and trivialize other modes of reasoning. Stanley (1978) has called this dominance by scientific and technical modes of reasoning "technicism." His ideas were discussed and applied to information policy in Chapter 4. In another time and place, a similar danger might be faced if organizational/economic knowledge (Wolfe, 1989) or cultural knowledge predominated and trivialized other types. Since technicism is recognized as the current problem, the focus here shall be on what technicism entails and why it is potentially dangerous for the design and evaluation of information policy.

TECHNICISM AS THE CURRENT PREDOMINANT MODE OF REASONING IN INFORMATION POLICY

If one looks at medical practice purely in terms of medical technology or applies technological modes of reasoning to the organizational and cultural spheres, this is what Stanley (1978) called *technicism*. Stanley defines technicism as:

> a state of mind that rests on an act of conceptual misuse, reflected in myriad linguistic ways, of scientific and technological modes of reasoning. This misuse results in the illegitimate extension of scientific and technological reasoning to the point of imperial dominance over all other interpretations of human existence. (p. 12)

Technicism "is basically a species of cognitive conquest" (Ibid., p. 14). Essentially it is reductionism.

In applying Stanley's mode of reasoning to information policy, an analyst would justifiably condemn as reductionism any attempt to see and evaluate information policy solely in terms of economics or law or the types of technology used in the carrying out of various policies, or the types of technology over which policy tries to exert control. A just and comprehensive evaluation must encompass all of these aspects, if an analyst wants to produce the best information policy possible. An analyst may get the organizational, economic aspects right, but ignore cultural values that can be the deciding factor in the success or failure of a policy. Errors of this kind were explored from different perspectives in Chapters 1–4.

In the area of information policy evaluation, an analyst could cite as technicism any attempt to claim assured predictability of actions taken to achieve specific ends. Scientific and technical modes of reasoning allow scientists to predict with varying degrees of reliability and validity the results of actions taken *in the natural world*. But if it is remembered that, with information policy, one is not restricted to natural phenomena or technological equipment, then scientific prediction, as discussed in Chapters 4 and 6, cannot be a reliable enterprise. Analysts and policy makers have to guard against conceiving the social world, the arena of public policy, as if it were a controlled experiment in a laboratory. For information policy, this means not reducing information policy problems to problems of technology or economics or law alone.

Stanley (1978) mentions four caveats that may help us to avoid technicism in the evaluation and design of information policies. His first caveat is that one should not view the social world as "an object with its own 'laws' (or 'lawlike regularities')" (p. 15). One must allow for the "creative agency of persons." He warns that:

> Humanistic criticism implies that when social science assumptions (of determinism and the prediction-control test of truth) are extended beyond the bounds of science, they can become the controlling assumptions of other domains of the social world. To the extent that this is so, a technicist culture is in the making. (Ibid., p. 15)

A second caveat begins with the observation that:

> In science, 'goals' are sought-after answers to research questions that take the form of specified and operationalized statements about measurable phenomena. At least in principle, scientific methods stress the virtues of technologically clear steps of inquiry. (Ibid.)

If this mode of reasoning is carried over to all domains of human activity, then "goals are thought of as reducible to technological means of address" (Ibid.). This is a sign that technicism is beginning to take hold.

His third caveat concerns problems. "To the extent that science addresses the notion of problems at all, it tends to regard them in principle as solvable. Problems are assumed to arise from the operation of knowable laws of physical nature" (Ibid.). If analysts apply this conception of problems to the social sphere, they are in danger of forgetting that "the word 'problems' also applies to experiences that are not entirely conceivable, and certainly not always solvable, in this technological manner" (Ibid.). Is there now, for example, a danger of doing this in the realm of scientific and technical information policy?

Finally, his fourth caveat is that an analyst should not make "policy implementation synonymous with applied science" (Ibid.). He does not mean to imply from this that "the honored desire to apply reason in the form of knowledge to one's problems must eventuate in technicism. This is true by definition so long as the concept of reason (and of knowledge) is officially limited to the sense of scientific expertise" (Ibid., p. 16). The word "knowledge," as used in this book, has always extended beyond the limits of scientific expertise to include organizational and cultural knowledge.

SPECIALIZED KNOWLEDGE AS THE BACKBONE OF INFORMATION POLICY

The case study of the SATCOM report illustrated a framework for evaluation that utilized content analysis in order to discover several salient parts of policy recommendations or policy tasks. Along with this framework of actor, goal, and rationale, and the various desiderata for evaluation discussed in Chapter 4, must be Stanley's caveats about goals and problems. To design a policy well or to evaluate one that has already been implemented, an analyst must start with these basic parts of the policy task in order to sort out actors, goals, and reasons for the actions recommended. Analysts must always question, however, whether the goal set is a legitimate one, given the policy task at hand. Moreover, they must be able to discover the appropriate knowledge and mode of reasoning that will be applicable to judge the rationale of any policy task. The case study took a preliminary step towards this end in trying to ascertain the relevance of information science to selected SATCOM policy tasks. But this process alone will never guarantee the soundness of an information policy. Based on the outcome of that case study, the next section suggests where information science (or a related complex of disciplines) has to develop in order to be up to the task of evaluating information policy.

DESIDERATA FOR THE DEVELOPMENT OF INFORMATION SCIENCE AS THE HOME DISCIPLINE FOR INFORMATION POLICY

The description of information science depicted below is indebted to Pacey's concept of the three areas of technology practice: the technical aspect, the organizational aspect, and the cultural aspect. Information science should also

be similarly regarded as a humanistically based discipline that has a technological component, an organizational component, and a cultural component.
These components exist on both a microinformatic level and a macroinformatic level. Figure 9–2 depicts this more clearly.

This figure requires some explanation. Across the horizontal border of the
figure are the headings for three aspects of information practice: cultural,
organizational, and technical. Along the vertical border is the microinformatic/
macroinformatic dichotomy, which was described previously. Macroinformatics refers to

> the study of information transfer phenomena 'in the large,' without reference to
> the views or interests of particular segments or elements of society, while
> microinformatics would imply the study of information transfer phenomena
> from the varying perspectives of these particular segments or elements. (Lan-
> caster and Burger, 1990, p. 153)

The phrases appearing in the grid are a preliminary attempt both to identify
areas of research for a developed information discipline and to describe
aspects of information policy making and policy research that should be
considered in the evaluation and design of information policies.

For example, in evaluating a scientific and technical information policy, an
analyst should not only be concerned with the reliability of a system of
distribution and access of scientific and technical information, but must also
see how such a proposed policy affects or would be affected by the users
themselves, personal scientific activity, broadly held societal values, macro-

Figure 9–2. Aspects of Microinformatics and Macroinformatics

	CULTURAL ASPECT	ORGANIZATIONAL ASPECT	TECHNICAL ASPECT
MICRO-INFORM-ATICS	creativity, personal values, awareness, belief in progress, image of homo informaticus	information microeconomics, institutional administrative structure, personnel policies, user activity, maintenance activity	informaion technology, retrieval and storage activity, system reliability, representation of information
MACRO-INFORM-ATICS	societal values and paradigms. belief in efficacy of human intervention; wisdom	professional societies, type of political institutional design, information macroeconomic policies, legal constraints	bibliometrics, telecommunications, broadcasting media

economic policies, etc. The application of these suggestions admittedly is best discussed against the background of a specific policy.

At the present time analysts do not seem to be making these suggested connections between relevant areas of information science or between areas of information science and information policy. Lancaster and Burger suggested that:

> An examination of a sample of the literature of information science shows that the great majority of all papers deal exclusively with topics in microinformatics. The field is long on practice and short on theory. Without vigorous and directed theoretical development, information science will dry up or become concerned solely with information technology. Unlike sociologists and economists, for example, information scientists seem not even to know what the central problems of information science are: the allocation of scarce information resources, the efficient utilization of information resources, maximizing of information use, or some principle that is operative but not yet sufficiently articulated?
>
> In the making of economic and social policies, economists, and sociologists have been able to make valuable contributions because the disciplines of sociology and economics have developed theoretical models to deal with phenomena in the aggregate and the interrelationships between the macro phenomena and the micro phenomena have been explored. The discipline of information science, on the other hand, seems not to have made an explicit distinction between microinformatics and macroinformatics and to have devoted almost all of its energy to microinformatic detail. (1990, pp. 155–156)

In terms of Figure 9-2, the focus has been predominantly on the technical aspect of microinformatics. The operational boundaries of information science, as elicited from the experienced practitioners of the case study also support this contention. Analysts need to expand their vision of the discipline to include these other aspects and other levels if they are to engage in a humanistically based discipline that affords dignity and justice to human beings. This expansion will also enable them to deal with information policy more effectively, both in its design and its evaluation.

How can this task be accomplished? The first step is the fostering of an awareness on the part of information policy makers and information professionals that they must avoid reductionism of all kinds, especially technicism. Reductionism, however convenient for evaluating efficiency, ignores the reasons for specific policy tasks and the specialized knowledge that can assess the validity of these reasons from several perspectives.

This brings us full circle to the previously asked questions: Will information policies have the effects policy makers intend? Will the policies have effects that analysts and policy makers would rather not think about? and Will specific disciplines yield answers to these policies? Furthermore, the schema of microinformatics and macroinformatics might prompt the following question: Will analysts and policy makers be able to encompass the three suggested aspects of information science on the micro- and macrolevels in their thinking, designing, and evaluating of information policies? The path that we now seem to be following towards answering these questions does not offer much hope. It only offers the possibility of social engineering and the diminution of the dignity and potential of human action in human affairs. OMB Circular A-130, with its concept of "information life cycle," provides an example of a current policy that ignores cultural effects and illustrates the predominance of technicism in designing and evaluating information policy.

In order to demonstrate the concrete relevance of these questions to current information policy, the following section will examine OMB's concept of "information life cycle," offer a critique of it, question whether "information life cycle" is applicable to all types of information, and suggest its ties to the evolving concept of information resources management. In conclusion, it will be shown that suggestions made in this chapter will apply to a revision of OMB's Circular A-130.

RELEVANCE TO CURRENT INFORMATION POLICY

OMB's Concept of "Information Life Cycle"

The concept of "information life cycle" is briefly described in Appendix IV to Office of Management and Budget's (1985) Circular No. A-130, "The Management of Federal Information Resources:"

> Good management and the requirement of practical utility dictate that agencies must plan from the outset for the steps in the information life cycle. The Act [Paperwork Reduction Act] also stipulates that agencies must "formulate plans for tabulating the information in a manner which will enhance its usefulness to other agencies and to the public" (44 U.S.C. 3507 (a)(1)(C)). When creating or collecting information, agencies must plan how they will process and transmit the information, how they will use it, what provisions they will make for access to it, whether and how they will disseminate it, how they will store it, and finally, how the information will ultimately be disposed of. While agencies cannot at the outset achieve absolute certitude in planning for each of these processes, the requirement for information resources planning is clearly contained in the Act (44 U.S.C. 3506(c)(1)), and the absence of adequate planning is sufficient not to create or collect information in the first place.

Sprehe (1987b, p. 194) elaborates on the information life cycle and the need for Federal agency planning. He points to OMB Bulletin 86–12, "Federal Information Systems and Technology Planning," which discusses the Circular's emphasis on information technology planning processes. It requires agencies to prepare strategic plans. He explains:

> Planning for information management—as contrasted with information systems and technology management—entails systematic consideration of the *information life cycle* [emphasis added]: creation, collection, processing, transmission, dissemination, use, storage, and disposition of information. As Appendix IV to the Circular points out:
>
>> If agencies do contemplate disseminating particular information, they should plan for its dissemination when creating or collecting the information. The focus of information dissemination plans should be on elevating to a policy level decisions regarding the agency's positive obligations to disseminate information and ensuring that the agency discharges the obligations in the most efficient, effective, and economical manner.
>
> Planning translates organizational missions into specific goals and, in turn, into quantifiable objectives. Planning then develops the means to achieve objectives, chooses among alternatives, and implements the chosen plans, monitoring progress and modifying plans as objectives and the environment change. (Ibid., pp. 194–195)

The concept of information life cycle, which originated among archivists in the 1940s (Bass and Plocher, 1989, p. 20), is the key to OMB's information policy activities. On the surface, the concept is a logical one. Federal agencies should not indiscriminately create, gather, and disseminate information unless it can fit into predictable patterns of use. The logic is flawed, however. The language used in the aforementioned OMB policy statements seems to assume that the use of information and, hence, its entire life cycle, is predictable. The entire concept of life cycle is a biological term applied to information flows. It carries with it connotations of lawlike regularities and predictability. Strategic planning, used in conjunction with "information life cycle," reduces the social world to a manageable board game that can be won if only the right strategy is used. In this case, Stanley's (1978) warnings, given above, are relevant: Analysts should not

(1) view the social world as "an object with its own 'laws' (or 'lawlike regularities')" (Ibid., p. 15);
(2) think of goals as "reducible to technological means of address" (Ibid.);
(3) regard all problems as solvable, since from a technicist viewpoint "problems are assumed to arise from the operation of knowable laws of physical nature" (Ibid.); and
(4) make "policy implementation synonymous with applied science" (Ibid.)

A broader view of the public enterprise that acknowledges the unpredictability of social events and that attempts to protect human welfare and culture first is needed. The efficient management of information resources is also important, but policy makers should not let it predominate over other, often less tangible, needs and goals.

The OMB policy, as embodied in Circular A-130, using as it does the concept of "information life cycle," is an example of technicism in practice. Reduction of information to a unidimensional economic commodity that can be quantified and managed in a precise way is a species of cognitive conquest, to use Stanley's phrase. Circular A-130 is simply not appropriate to the problem confronted. It does not put the welfare of human beings first, and it is carried out on a scale too grand to achieve its desired end.

Given these criticisms, it is not surprising that the results of the present OMB policy have come under fire from several quarters (Bass and Plocher, 1989; Hernon, 1986; Plocher, 1988). While their comments have been useful, the criticism made here is broader and more theoretical. Here it is argued that the philosophical foundations of the policy are objectionable. The reductionist mode of reasoning, present in the concept of information life cycle and the policy implications drawn from it, exclude other considerations, such as the allowance for additional value to be placed on the information as it is used, changes in the utility of the information as it is used, and the need for information as the lifeblood of democracy. The mode of thinking reflected in Circular A-130 makes this and similar considerations extraneous to the argument since they cannot be quantified and fit into the existing system of management and control.

The primary goal of information policy, as seen by OMB in 1985, was a quantifiable, economic, one. The use of the Information Collection Budget (Rubenstein, 1990; Bass and Plocher, 1989, pp. 13–14) by OMB to control the amount of Federal agency information to be disseminated treats information as a burden to be eliminated and reduced. A democratic view of information rejects this principle, arguing instead that information dissemination is a positive force, often unmeasurable, in maintaining a democratic society (Bass and Plocher, 1989, pp. 32–33).

The information life cycle concept focuses primarily on the microinformatic organizational area with some reliance also on the microinformatic technical aspect. Furthermore, the knowledge used in supporting the program is only appropriate to these areas. It ignores completely the cultural aspect of micro and macroinformatics. In light of the previous discussion of Stanley's analysis, the concept and Circular A-130 may legitimately be called technicist.

Fortunately, since 1987, there is evidence that OMB and its Office of Information and Regulatory Affairs (OIRA), has begun to move away from the unambivalently technicist position reflected in Circular A-130. It has done this, in part, because of the criticism voiced publically about A-130, but also in part because Circular A-130 did not emcompass the broad range of information policy problems confronted by the Federal government. One of these problems, the electronic dissemination of information, was not originally addressed by the Circular. Furthermore, OIRA apparently concluded that information was not a unidimensional economic commodity that could be controlled in a clear-cut, unambiguous way. Information policy actions had effects not measurable by econometric methods. This is not to say that the current thinking at OMB regarding information policy is without fault, but it does indicate a positive move away from a unidimensional technicist approach to this highly complex problem. A brief examination of the evolution of OMB's position statements will reveal the more positive direction in which it is moving.

Evolution of Thinking about Federal Information Policy at OMB

On August 7, 1987, OMB issued "Policy Guidance on Electronic Collection of Information" (U.S. Office of Management and Budget, 1987). The notice was issued in response to the 1986 amendment to the Paperwork Reduction Act (44 U.S.C. Chapter 35, see especially 3501 (5)) that mandated improved management of Federal information resources and suggested the increased use of information technology to reduce the information-processing burden of both the Federal government and the public, and to improve the efficiency and effectiveness of Federal agency operations. Some examples where information technology would be of help were increasing the automation of the submission of Medicare forms, encouraging electronic submission of Shippers Export Declaration (SED) Reports, and broadening the use of electronic filing of individual income tax returns and similar programs. The purpose of the policy guidance offered by OMB had two components. First, it was designed "to cause agencies systematically to take account of potential management efficiences derivable from electronic information collection, and second to ensure that agencies consider the major legal and policy issues that arise in connection with such collection" (U.S. Office of Management and Budget, 1987, p. 29,456).

Statements made in this notice indicate that the notion of burden was changing and was not now seen exclusively in economic terms. The information resources management concepts introduced placed more emphasis on dissemination and shifted the focus from immediate economic benefit to a longer-range perspective. This shift in focus is evidenced by the requirement to design for future dissemination in the collection phase of any information collection program.

Within a year and a half the further evolution of these positions was evident in another notice, "Advance Notice of Further Policy Development on Dissemination of Information" (U.S. Office of Management and Budget, 1989a). In a 1986 amendment to the Paperwork Reduction Act, the word "dissemination" was introduced in several places in the law. In response to the new wording of the amendment, OMB sought comment on the expansion of Circular A-130 to cover information dissemination policy. In conjunction with this further development, OMB proposed incorporating OMB Circular A-3, "Government Publications." OMB also sought the answers to the questions concerning the comprehensiveness and enforceability of the proposed policy.

In the "Analysis of Policy" section of this notice, OMB tried to counter the previous interpretations of A-130 concerning the public use of information by stating:

> Efficient, effective, and economical dissemination does not translate into diminishing or limiting the flow of information from the agency to the public. To the contrary, good management of information resources should result in more useful information flowing with greater facility to the public, at less cost to the taxpayer. (Ibid., p. 214)

It further buttressed this position by quoting part of the Paperwork Reduction Act (44 U.S.C. 3501): "A basic purpose of the Paperwork Reduction Act is 'to maximize the usefulness of information collected, maintained, and disseminated by the Federal Government'" (Ibid., p. 217).

In order to effect these ends, the proposed policy presented several requirements, some more specific and concrete than others. Among them were:

- Manage information "so as to maximize efficient and effective performance of agency functions, maximize the usefulness of government information, and minimize the cost to the Federal government;"
- Implement a management control system for all information dissemination products;
- Expend funds only for periodicals that provide information necessary for conducting public business;
- Determine the significance of information dissemination products and provide mechanisms for notifying the public if and when these products would be discontinued; and
- Avoid disseminating products that would lead to unfair competition with the private sector. (Ibid.)

Special care was taken to describe the conditions under which user charges would be considered appropriate and under which circumstances agencies should rely on the private sector for the dissemination of certain information. "In effect," the notice said, "agencies should prefer to 'wholesale' government information and leave 'retail,' value-added functions to the private sector, especially when they know that the private sector is ready and able to perform the value-added functions" (Ibid., p. 217).

The response to this notice led to another notice in June 15, 1989, which withdrew the proposed policy. What had gone wrong?

The most common observation was that the notice of January, 1989 and Circular A-130 "were heavily biased, concentrating so much on private sector prerogatives that OMB had failed to elaborate a positive role for Federal agencies in the dissemination of information, even in situations where dissemination of information was basic to agencies' missions" (U.S. Office of Management and Budget, 1989b, p. 25,554). Commentators generally did agree, however, that sharing responsibility for dissemination between the private and public sectors was both "good public policy and good economics" (Ibid.).

Several other issues were addressed in the June, 1989 notice, including the incorporation of OMB Circular A-3 into Circular A-130, an exhortation that individual states take a more active role in the dissemination of Federal information, objections to the proposed elimination of the exclusion of statistical periodicals from the reporting requirements of Circular A-130, and questions concerning user charges for information.

Perhaps the most interesting response to come from OMB was an unambiguous statement regarding its philosophy of information dissemination:

> OMB wishes to make clear that its fundamental philosophy is that government information is a public asset: that is, with the exception of national security matters and such other areas as may be prescribed by law, it is the obligation of government to make such information readily available to the public on equal terms to all citizens. (Ibid., p. 25,557)

This statement seems to indicate that the extreme technicist position, stated clearly in earlier policy documents, had been mollified. At issue, of course, are the principles by which OMB judges whether dissemination is good or worthwhile. Under earlier pronouncements, an economic calculus was the prevailing factor; in this latest statement, the stress has also shifted to the value and use of information, not simply the cost of production.

The January 1989 notice and the public reaction to it signaled a theoretical turning point, from the crude and imprecise measurements of information value typified by the Information Collection Budget and emphasis on burden, to a broader concern for the public good and an awareness that information resources management is a multifaceted, not a unifaceted affair. The subsequent June 1989 notice seemed to indicate a rejection by some of the information policy principles of the Reagan administration, most notably the privatization of government information resources. This notice indicated a resolve to go back to the drawing board of information policy formulation and produce a new draft policy that would overhaul information collection and information dissemination policy. That overhaul began with another solicitation for public comment on March 4, 1991.

The tone and content of the March 4, 1991 notice, "Advance Notice of Plans for Revision of OMB Circular no. A-130, Management of Federal Information Resources" (U.S. Office of Management and Budget, 1991), is notable for its openness to past critics and its plans for a thorough revision of Circular A-130. The primary emphasis in the new A-130, the notice proclaims, will be on the revision of information dissemination policy. Elements that had appeared in earlier notices, such as the role of information technology in dissemination, a more positive relationship with users of information exemplified by the adequate notice provision (giving public notice prior to "creating, terminating, or making significant changes to major information products" (Ibid., p. 9027)), user charges, avoidance of monopolistic practices by the government, and the relationship between Federal and non-Federal dissemination of information, are all earmarked for treatment in the revised A-130.

Furthermore, OMB promised to treat new topics that arose in conjunction with responses to earlier notices. These new topics include the role of the states in Federal IRM, greater attention to records management and disposition, electronic collection of information, and, probably most importantly, strategic information resources planning and cost/benefit analysis. Although this may at first sound like a new idea, the practical outcome of such policy may be the same as it was when OMB used "information life cycle" as a tool to plan strategically and evaluate information programs.

Has progress been made with Federal information policy? Theoretically it has to some degree. The language used in the notices examined seems to indicate both an increased concern for the public good and the intelligent use of Federal information resources to further that end. There is also now evident a responsiveness to the private sector that was not manifest several years ago. In another sense the entire theoretical direction of the policy seems to be technicist. Although "information life cycle" is not used in these notices, phrases such as "strategic information resources planning and cost/benefit analysis" carries with it several of the trappings of technicism, such as the assumption that the problem is solvable and the implied exclusion of cultural factors in the formulation, design, and implementation of policy.

Furthermore, the Federal information policy envisioned by OMB is understandably circumscribed by OMB's obligations under the Paperwork Reduction Act, which lapsed in 1989, not to be reauthorized. This cannot be helped, but this entire book has been devoted to exploring other areas that need to be part of any national information policy. There is no indication, for example, that there are international implications to this policy, nor is there any indication that the policy is anything more than an internal governmental guideline for the efficient operation of Federal agencies. More shortsighted, however, is the inadequate evaluation of the policy. As it stands now, the Information Collection Budget (ICB) is the lone evaluative and control tool offered by OMB. There has not been adequate exploration of what is meant by *maximation of use*. What does this mean exactly? Can it be measured? What rationales are used to justify what is meant by maximization, other than the crude measure of burden hours? Will answers to these questions be applicable to the micro- and macrolevels?

To answer these questions, users of Federal information other than the agencies themselves must be involved in policy development and evaluation (McClure, Bishop, Doty and Fellows, 1988). This also has a bearing on the use of information technology, which has different objectives and configurations when used for internal control as opposed to meeting user needs.

Another area ignored in these notices is the different character of various types of information. For example, A-130 and its related notices do not distinguish scientific and technical information (STI) from other types of information. The uses of STI probably do not conform to the information life cycle scheme. The use of STI is different from that of commercial information or artistic information. How these types of information differ is not something that can be explored here, but it does raise the question of the adequacy of a policy that fails to distinguish between types of information and assumes that these different types can be treated and their use "strategically planned" in a common way. Thus, in spite of the theoretical progress made, it is unclear that concrete progress is manifest in the day to day implementation by OMB of Federal information policy.

FUTURE NEEDS

What is clearly needed is a historical study of the formulation, implementation, and evaluation of Circular A-130 and other facets of national information policy. Only by examining in detail what was done with a specific written policy can policy analysts, legislators, and private citizens determine what the concrete effects of the current policy have been or what effects future policies are likely to have. Furthermore, in conjunction with such a historical study, an analysis of the policy documents and other formal statements should be carried out, along the lines suggested by the analysis of the SATCOM Report. Only after such study can a comprehensive and culturally sensitive policy be made.

This task is not an easy one, nor one that can be carried out quickly. It is frought with difficulties and complexities. The effects of existing information policy, however, are now only partially understood and often viewed through the prism of information resources management and information life cycle. These concepts are helpful, but they have a way of shifting the focus from the more subtle, but much more important, effects of information control. This is the problem this country faces—not the efficient management of information resources, for this is stating the problem too narrowly. Instead the problem is that of the unavoidable control of information resources. How available information is controlled, who controls it, and why it is controlled in the way it is are questions whose answers will determine the health of the body politic in the years to come.

REFERENCES

Abate, D. (1989). Essential features of a national information and documentation network: A PADIS perspective. In O.C. Mascarenhas.(Ed.), *Establishment of a National Information and Documentation Network in Tanzania: Papers of the Seminar Held in Dar es Salaam 16 to 24 February, 1989.* Bonn: Deutsche Stiftung fur internationale Entwicklung, Zentralstelle fur Erziehung, Wissenschaft und Dokumentation, 76–84.

Adkinson, Burton W. (1978). *Two Centuries of Federal Information.* Stroudsburg, PA: Dowden, Hutchinson & Ross.

Aines, Andrew A. (1984, May). A visit to the wasteland of federal scientific and technical information policy. *Journal of the American Society for Information Science, 35,* 179–184.

Albaek, Erik (1989–1990, Winter). Policy evaluation: Design and utilization. *Knowledge in Society, 2,* 6–19.

Allen, T. J. (1985). *Managing the Flow of Technology: Technology Transfer and the Dissemination of Technological Information within the R&D Organization.* Cambridge, MA: MIT Press.

Anderson, D. (1974). *Universal Bibliographic Control: A Long Term Policy: A Plan for Action.* Pullach/Munchen: Verlag Dokumentation.

Andren, Gunnar (1981). Reliability and content analysis. In Karl Erik Rosengren (Ed.), *Advances in Content Analysis.* Beverly Hills, CA: Sage Publications, 43–67.

Andrews, E. L. (1990, August 22). Tiny Tonga seeks satellite empire in Space. *The New York Times,* A1 and C17.

Ashoor, M. S. (1988). Bibliographic networking in the Gulf region. In Ziauddin Sardar (Ed.), *Building Information Systems in the Islamic World.* London: Mansell Publishing Ltd., 116–124.

Ballard, Steve (1987). Federal Science and Information Policies: An Overview. In P. Hernon and C. McClure (Eds.), *Federal Information Policies in the 1980s: Conflicts and Issues.* Norwood, NJ: Ablex, 195–225.

Bass, Gary and Plocher, David (1989). *Strengthening Federal Information Policy: Opportunities and Realities at OMB.* Washington, DC: Benton Foundation (Benton Foundation Project on Communications & Information Policy Options; 6)

Beniger, J. R. (1986). *The Control Revolution: Technological and Economic Origins of the Information Society.* Cambridge, MA: Harvard University Press.

175

Bezold, Clement and Olson, Robert. (1986). *The Information Millennium: Alternative Futures: A Report for the Information Industry Association.* Washington, DC: Information Industry Association.

Bikson, Tora K., Quint, Barbara E., and Johnson, Leland L. (1984). *Scientific and Technical Information Transfer: Issues and Options.* (Rand Note 2131). Santa Monica, CA: Rand Corporation .

Bishop, Ann and Fellows, Maureen O'Neill. (1989). Descriptive analysis of major federal scientific and technical information policy studies. In C. McClure and P. Hernon (Eds.), *U.S. Scientific and Technical Information (STI) Policies: Views and Perspectives.* Norwood, NJ: Ablex, 3–55.

Borgmann, A. (1984). *Technology and the Character of Contemporary Life: A Philosophical Inquiry.* Chicago: University of Chicago Press.

Boyce, Bert R. and Kraft, Donald H. (1985). Principles and theories in information science. *Annual Review of Information Science and Technology, 20,* 153–78.

Braman, Sandra. (1988) *Information Policy and the United States Supreme Court.* Ph.D. Thesis. University of Minnesota.

Browne, Malcolm W. (1990, November 22). 57% of U.S. math doctorates going to foreigners. *The New York Times,* A13.

Burger, Robert H. (1986). The analysis of information policy. In Robert H. Burger (Ed.), *Privacy, Secrecy, and National Information Policy (Library Trends,* 35, (1), Summer). Urbana, IL: University of Illinois Graduate School of Library and Information Science, 171–182.

———. (1988). *The Evaluation of Information Policy: A Case Study Using the SATCOM Report.* Ph.D. Thesis. University of Illinois at Urbana- Champaign. (UMI Order No. 8908634).

Burkett, J. (1983). *Library and Information Networks in Western Europe.* London: Aslib.

Bushkin, A. A. and Yurow, J. H. (1981). The foundations of United States information policy. In Hans-Peter Gassmann (Ed.), *Information, Computer and Communications Policies for the 80's: An OECD Report: Proceedings of the High Level Conference on Information, Computer and Communications Policies for the 80's, Paris, 6th-8th October, 1980,* Amsterdam: North-Holland, 215–226.

Case, Donald. (1985). Information science in the eighties. *Library Science Annual, 1,* 46–50.

Chartrand, Robert Lee. (1986a). Legislating information policy. *Bulletin of the American Society for Information Science, 12,* (5), 10.

———. (1986b). Policy imperatives in the information age. *Bulletin of the American Society for Information Science, 12,* (3), 6 and 9.

——— and Chalk, Rosemary A. (1975). *Federal Management of Scientific and Technical Information (STINFO) Activities: The Role of the National Science Foundation.* Washington, D.C., Congressional Research Service, (issued as Committee Print, Committee on Labor and Public Welfare, United States Senate, 94th Congress, 1st Session). Washington, D.C.: GPO.

Computer-Readable Databases: A Directory and Data Sourcebook, 5th edition. (1989). Detroit: Gale Research, Inc.

Cronin, B. (1987). Telematics and retribalisation. *Aslib Proceedings, 39,* 87–95.

Denenberg, R. (1988). Standard network interconnection protocols. In Judith G. Fenley and Beacher Wiggins (Eds.), *The Linked Systems Project: A Networking Tool for Libraries* (OCLC Library, Information, and Computer Science Series, 6). Dublin, OH: OCLC Online Computer Library Center, 19–49.

Dizard, Wilson P. (1982). *The Coming Information Age: An Overview of Technology, Economics and Politics.* New York: Longman.

Doty, Philip and Erdelez, Sanda. (1989). Overview and analysis of selected federal scientific and technical information (STI) policy instruments, 1945–1987. In C. McClure and P. Hernon (Eds.), *U.S. Scientific and Technical Information (STI) Policies: Views and Perspectives.* Norwood, NJ: Ablex, 56–83.

Durkheim, E. (1893). *The Division of Labor in Society.* Translated by George Simpson. New York: Free Press.

Ellsworth, John W. (1965, December). Rationality and campaigning: A content analysis of the 1960 presidential campaign debates. *The Western Political Quarterly, 18* (4), 794–802.

Ellul, Jacques. (1964). *The Technological Society.* Translated by John Wilkinson. New York: Alfred A. Knopf.

Eres, B. K. (1989). International information issues. *Annual Review of Information Science and Technology, 24,* 3–32.

Feldman, Martha S. (1989). *Order Without Design: Information Production and Policy Making.* Stanford, CA: Stanford University Press.

Flynn, Roger R. (1987). *An Introduction to Information Science.* New York: Marcel Dekker.

Foskett, D. J. (1970, December). Informatics . *Journal of Documentation, 26,* 340–369.

Garvey, W.D. (1979). *Communication: The Essence of Science.* New York: Pergamon Press.

Gavrilova, S. A., Kuskov, A. N., and Tyshkevich, N. I. (1981). *Mezhdunarodnaia sistema nauchnoi i tekhnicheskoi informatsii stran-chlenov SEV: Lektsii (The International System of Scientific and Technical Information: Lectures).* Moscow: MTSNTI.

Gilyarevsky, R.S. (1990). The contribution of informatics to the solution of semantic problems. In D.J. Foskett (Ed.), *The Information Environment: A World View. Studies in Honour of Professor A.I. Mikhailov.* Amsterdam: Elsevier (FID 685), 159–166.

Goodman, S. E. (1988). Information societies. In Richard F. Staar (Ed.), *The Future Information Revolution in the USSR.* New York: Crane Russack, 11–18.

Gould, Stephen B. (1986). Secrecy: its role in national scientific and technical information policy. In Robert H. Burger (Ed.), *Privacy, Secrecy and National Information Policy (Library Trends, 35* (1), Summer). Urbana, IL: University of Illinois Graduate School of Library and Information Science, 61–81.

Gray, J. (1988). National policy issues. In O. Mohamed (Ed.), *Formulating a National Policy for Library and Information Services: The Malaysian Experience.* London: Mansell Publishing, 110–116.

Hayes, Robert M. (1985). Introduction. In R.M. Hayes (Ed.), *Libraries and the Information Economy of California.* Los Angeles: University of California at Los Angeles, 1–49.

Heim, Kathleen M. (1986). National information policy and a mandate for oversight by the information professions. *Government Publications Review, 13,* 21–37.

Herner, Saul. (1984). Brief history of information science. *Journal of the American Society for Information Science, 35,* 157–63.

Hernon, Peter. (1986). The management of United States government information resources: An assessment of OMB Circular A-130. *Government Information Quarterly, 3,* 279–290.

———. (1989). Protection of U.S. STI under the Reagan Administration. In Peter Hernon and Charles McClure (Eds.), *U.S. Scientific and Technical Information (STI) Policies:Views and Perspectives.* Norwood, NJ: Ablex, 87–108.

——— and McClure, Charles R. (1987). *Federal Information Policies in the 1980s: Conflicts and Issues.* Norwood, NJ: Ablex.

Houser, Lloyd. (1988). A conceptual analysis of information science. *Library and Information Science Research, 10,* 3–34.

Information Policy for the 1980s: Proceedings of the Eusidic Conference, Copthorne, U.K., 3–5 October 1978. (1979). Oxford: Learned Information.

Intergovernmental Conference for the Establishment of a World Science Information System. (1971). *UNISIST: Final Report, Paris 4–8 October 1971.* Paris: Unesco.

Jacob, M.E.L. and Rings, D.L. (1986). National and international information policies. *Library Trends, 35,* (1), 119–161.

Jenkins, W. I. (1978). *Policy Analysis: A Political and Organizational Perspective.* New York: St. Martin's Press.

Judy, R. W., and Clough, V. L. (1989). *The Information Age and Soviet Society.* New York: Hudson Institute.

Kamaruddin, A. R. (1988). Islam-Net: The development of an Islamic information network. In

Ziauddin Sardar (Ed.), *Building Information Systems in the Islamic World*. London: Mansell Publishing, 142–155.

Kashlev, I. B. (1988). *Informatsionnyi vzryv: mezhdunarodnyi aspekt (The Information Explosion: International Aspect)*. Moscow: Mezhdunarodnye otnosheniia.

Kelman, Steven. (1987). *Making Public Policy: A Hopeful View of American Government*. New York: Basic Books.

Keren, C. (1984). On information science. *Journal of the American Society for Information Science, 35*, 137.

Kochen, Manfred.(1983). Library science and information science: Broad or narrow? In Fritz Machlup and Una Mansfield (Eds.), *The Study of Information: Disciplinary Messages*, New York: John Wiley, 371–377.

Lancaster, F.W. and Burger, Robert H. (1990). Macroinformatics, microinformatics and information policy. In D.J. Foskett (Ed.), *The Information Environment: A World View. Studies in Honour of Professor A.I. Mikhailov*. Amsterdam: Elsevier (FID 685), 149–158.

Langlois, Richard N. (1983). Systems theory, knowledge and the social sciences. In Fritz Machlup and Una Mansfield (Eds.), *The Study of Information: Interdisciplinary Messages*. New York: John Wiley, 581–600.

Le Garff, A. (1982). *Dictionnaire de l'informatique*. Paris: Presses universitaires de France.

Lesk, M. E. (1970, September). [Review of SATCOM] *Computing Reviews, 11*, 513.

Lilley, Dorothy B. and Trice, Ronald W. (1989). *A History of Information Science: 1945–1985*. San Diego: Academic Press.

Lim, E. H. -T. (1989). Networking activities in Southeast Asia. *Information Services & Use, 9*, 5–32.

Lindblom, Charles E. (1980). *The Policy-Making Process*. 2nd ed. Englewood Cliffs, NJ. Prentice-Hall.

Lindkvist, Kent (1981). Approaches to textual analysis. In Karl Erik Rosengren (Ed.), *Advances in Content Analysis*. Beverly Hills, CA: Sage Publications, 23–41.

Line, M. & Vickers, S. (1983). *Universal Availability of Publications (UAP): A Programme to Improve the National and International Provision and Supply of Publications (IFLA Publications 25)*. Munich: K.G. Saur.

Linowes, David F. and Colin Bennett (1986). Privacy: Its role in federal government information policy. In Robert H. Burger (Ed.), *Privacy, Secrecy and National Information Policy (Library Trends, 35*,1). Urbana, IL: University of Illinois Graduate School of Library and Information Science, 19–42.

Linstone, Harold A. and Turoff, Murray. (1975). *The Delphi Method: Techniques and Applications*. Reading, MA: Addison-Wesley.

Losee, Robert M., Jr. (1990). *The Science of Information: Measurement and Applications*. San Diego: Academic Press.

Lowi, Theodore J. (1964, July). American business, public policy, case studies, and political theory. *World Politics, 16*, 677–715.

——————. (1972, July/August). Four systems of policy, politics and choice. *Public Administration Review, 32*, 298–310.

MacBride, S. (1980). *Many Voices, One World: Towards a New More Just and More Efficient World Information and Communication Order*. New York: Unipub.

McClure, Charles R. (1989). *Testimony before the U.S. House of Representatives, Committee on Science, Space and Technology, Subcommittee on Science, Research and Technology. Hearings on Federal Scientific and Technical Information Policy, October 12, 1989*. Washington, DC: GPO.

——————, Bishop, Ann, Doty, Philip, and Fellows, Maureen O'Neill. (1988). Federal scientific and technical information (STI) policies and the management of information technology for the dissemination of STI. *ASIS '88: Proceedings of the 51st ASIS Annual Meeting, 25*. Medford, NJ: Learned Information.

——————, Hernon, Peter, and Relyea, Harold C. (1989). *United States Government Information Policies: Views and Perspectives*. Norwood, NJ: Ablex.

Machlup, Fritz, and Mansfield, Una. (1983). Cultural diversity in studies of information. In F. Machlup and U. Mansfield (Eds.), *The Study of Information: Interdisciplinary Messages*. New York: John Wiley, 3–56.

McIntosh, Toby J. (1990). *Federal Information in the Electronic Age: Policy Issues for the 1990s.* Washington, DC: Bureau of National Affairs

Mahon, B. (1989). Developments in European information policy. *Perspectives in Information Management, 1,* 63–87.

Mandeville, Thomas. (1987). An international comparison. In T. Barr (Ed.), *Challenges and Change: Australia's Information Society.* Oxford: Oxford University Press in association with the Commission for the Future, 30–37.

Markoff, John. (1991, January 21). A fresh eye on the environment: The supercomputer assesses changes. *The New York Times,* C1–C2.

Marney, Carlisle. (1961). *Structures of Prejudice.* New York: Abingdon Press.

Masmoudi, M. (1979). The new information order. *Journal of Communication, 29,* 172–185.

Mason, Marilyn Gell. (1983). *The Federal Role in Library and Information Services.* White Plains, NY: Knowledge Industry Publications.

Mikhailov, A.I., Chernyi, A.I. and Giliarevskii, R.S. (1984). *Scientific Communications and Informatics,* translated by Robert H. Burger. Arlington, VA: Information Resources Press.

Milevski, Sandra. (1986). Information policy through public laws of the 95th-98th Congresses. *Proceedings of the ASIS Annual Meeting, 23,* 211–219.

Mohamed, Oli (Ed.). (1988). *Formulating a National Policy for Library and Information Services: The Malaysian Experience.* London: Mansell.

Moohan, G., Morton, E., Rimmer, S., Romano, G., and Burton, P. F. (1988). Transborder data flow: A review of issues and policies. *Library Review, 37,* 27–37.

Mowlana, H. (1985). *International Flow of Information: A Global Report and Analysis (Reports and Papers on Mass Communication,* no. 99). Paris: Unesco.

Nagel, Stuart S. (1984). *Basic Literature in Policy Studies: A Comprehensive Bibliography.* Greenwich, CT: JAI Press.

Nakamura, Robert T. and Smallwood, Frank. (1980). *The Politics of Policy Implementation.* New York: St. Martin's Press.

National Commission on Libraries and Information Science. (1982). *Public Sector/Private Sector Interaction in Providing Information Services.* Washington, DC: GPO.

National Information Policy: A Report to the President of the United States. (1976). Submitted by the Staff of the Domestic Council Committee on the Right of Privacy. Washington, DC: National Commission for Libraries and Information Science.

Olsgaard, John N. (Ed.). (1989). *Principles and Applications of Information Science for Library Professionals.* Chicago: American Library Association.

Pacey, Arnold. (1983). *The Culture of Technology.* Cambridge, MA: MIT Press.

Plocher, David. (1988). Institutional elements in OMB's control of government information. *Government Information Quarterly, 5,* 315–322.

Porat, Marc Uri, with the assistance of Michael Rogers Rubin. (1977). *The Information Economy* (9 Vols.). Washington, DC: Office of Telecommunications.

Rahman, S. U. (1988). Information resource sharing and network projects. In Ziauddin Sardar (Ed.), *Building Information Systems in the Islamic World.* London: Mansell Publishing, 125–141.

Ravault, Rene-Jean. (1987). The ideology of the information age in a senseless world. In Jennifer Daryl Slack and Fred Fejes (Eds.), *The Ideology of the Information Age.* Norwood, NJ: Ablex, 178–199.

Rayward, W. Boyd. (1983a). Library and information sciences: Disciplinary differentiation, competition, and convergence. In Fritz Machlup and Una Mansfield (Eds.), *The Study of Information: Disciplinary Messages.* New York: John Wiley, 343–369.

_____. (1983b). Library science and information science: Together or apart? In Fritz Machlup and Una Mansfield (Eds.), *The Study of Information: Disciplinary Messages*. New York: John Wiley, 400–402.

Rein, M. (1976). *Social Science and Public Policy*. New York: Penguin Books.

Relyea, Harold C. (1986). Secrecy and national commercial information policy. In Robert H. Burger (Ed.), *Privacy, Secrecy and National Information Policy*. (*Library Trends*, 35,1). Urbana, IL: University of Illinois Graduate School of Library and Information Science, 43–59.

———. (1985). *Striking a Balance: National Security and Scientific Freedom: First Discussions*. Washington, DC: American Association for the Advancement of Science, Committee on Scientific Freedom and Responsibility.

Richards, P. S. (1986). Government information policy in the German Democratic Republic. *Advances in Librarianship, 14*, 101–142.

Rosenberg, Victor. (1981). National information policies. *Annual Review of Information Science and Technology, 17*, 3–32.

——— and Whitney, Gretchen. (1986). *The Transfer of Scholarly, Scientific and Technical Information Between North and South America: Proceedings of a Conference*. Metuchen, NJ: Scarecrow Press.

Rossi, P. H. and Freeman, H. E. (1989). *Evaluation: A Systematic Approach*. 4th ed. Newbury Park, CA: Sage Publications.

Rubenstein, Gwen. (1990). The quantification of information: The paperwork budget and the birth of the burden hour. *Government Information Quarterly, 7* (1), 73–81.

Salisbury, Robert H. (1968). The analysis of public policy: A search for theories and roles. In Austin Ranney (Ed.), *Political Science and Public Policy*. Chicago: Markham Publishing, 151–175.

Salton, Gerard. (1985). A note about information science research. *Journal of the American Society for Information Science, 36*, 268–271.

Salvaggio, J. L. (1989). Is privacy possible in an information society? In Jerry L. Salvaggio (Ed.), *Information Society: Economic, Social, & Structural Issues*. Hillsdale, NJ: Lawrence Erlbaum, 115–130.

Sardar, Z. (1988). *Information and the Muslim World: A Strategy for the Twenty-First Century*. London: Mansell Publishing

Schact, Wendy. (1985). *Industrial Innovation: The Debate over Government Policy* Issue Brief (IB 4004). Washington, DC: Library of Congress, Congressional Research Service.

Schrader, Alvin M. (1983). *Toward a Theory of Library and Information Science*. Ph.D. Thesis. Indiana University. Ann Arbor, MI: University Microfilms.

———. (1984). In search of a name: Information science and its conceptual antecedents. *Library and Information Science Research, 6*, 227–271.

Scientific Communication and National Security. (1982). A report prepared by the Panel on Scientific Communication and National Security, Committee on Science, Engineering and Public Policy, National Academy of Science, National Academy of Engineering, Institute of Medicine. Washington, DC: National Academy Press.

Scientific and Technical Communication: A Pressing National Problem and Recommendations for Its Solution: A Report by the Committee on Scientific and Technical Communication of the National Academy of Sciences-National Academy of Engineering. (1969). Washington, DC: National Academy of Sciences.

Shannon, Claude and Weaver, Warren. (1949). *The Mathematical Theory of Communication*. Urbana, IL: University of Illinois Press.

Sherrod, John and McFarland, Marvin. (1976). National policy and dissemination of scientific and technical information. *Proceedings of the ASIS Annual Meeting, 13*, 73.

Sprehe, J. Timothy. (1987a). Implementing the government's new information policy. *Information Services & Use, 4/5*, 139–144.

———. (1987b). OMB Circular No. A-130, the management of federal information resources: Its origin and impact *Government Information Quarterly, 4*, 189–196.

Stanley, Manfred. (1978). *The Technological Conscience: Survival and Dignity in an Age of Expertise*. Chicago: University of Chicago Press.

Stevenson, Russell B., Jr. (1980). *Corporations and Information: Secrecy, Access, and Disclosure.* Baltimore: Johns Hopkins University Press.

The Story of the United States Patent and Trademark Office. (1981). Washington, DC: GPO.

Surprenant, T. T. (1987). Problems and trends in international and communication policies. *Information Processing & Management, 23*, 47–64.

Swanson, Rowena W. (1970, February). SATCOM in review. *Datamation, 16*, 98–99, 102–104.

Taylor, A. E. (1955). *Aristotle.* revised ed. New York: Dover Publications.

Toulmin, Stephen. (1990). *Cosmopolis: The Hidden Agenda of Modernity.* New York: The Free Press.

Trauth, Eileen. (1986) An integrative approach to information policy research. *Telecommunications Policy, 10*(1), 41–50.

Tutchings, Terrence R. (1979). *Rhetoric and Reality: Presidential Commissions and the Making of Public Policy.* (Westview Special Studies in Public Policy and Public Systems Management). Boulder, CO: Westview Press.

Unesco Statistical Yearbook. (1989). Paris: Unesco.

U.S. Office of Management and Budget. (1985, December 24). OMB Circular A-130: The management of federal information resources. *Federal Register, 50*, 52730–52751; *51* (Jan 6, 1986), 461.

———. (1987, August 7). Policy guidance on electronic collection of information. *Federal Register, 52*, 29454–29457.

——— (1989a, January 4). Advance notice of further policy development on dissemination of information. *Federal Register, 54*, 214–220.

——— (1989b, June 15). Second advance notice of further policy development on dissemination of information. *Federal Register, 54*, 25554–25559.

——— (1991, March 4). Advance notice of plans for revision of OMB Circular No. A-130, Management of federal information resources. *Federal Register, 56*, 9026–9028.

U.S. Congress. Office of Technology Assessment. (1986). *Intellectual Property Rights in an Age of Electronics and Information.* Washington, DC: Office of Technology Assessment.

———. (1988). *Informing the Nation: Federal Information Dissemination in an Electronic Age.* Washington, DC: Office of Technology Assessment.

———. (1990). *Helping America Compete: The Role of Federal Scientific & Technical Information.* Washington, DC: GPO.

Vogel, Ezra F. (1979). *Japan as Number One.* Cambridge, MA: Harvard University Press.

Von Laue, Theodore H. (1987). *The World Revolution of Westernization: The Twentieth Century in Global Perspective.* New York: Oxford University Press.

Wallace, Danny Paul. (1985). *The Relationship Between Journal Productivity and Obsolescence in a Subject Literature.* Ph.D. Thesis. University of Illinois at Urbana-Champaign.

Weinberg, Alvin. (1963). *Science, Government and Information: The Responsibilities of the Technical Community and the Government in the Transfer of Information.* Washington, DC: GPO.

Wiener, Norbert (1950). *The Human Use of Human Beings.* Boston: Houghton Mifflin.

Wijasuriya, D. E. K. (1988). Considerations for a national policy. In O. Mohamed (Ed.), *Formulating a National Policy for Library and Information Services: The Malaysian Experience.* London: Mansell Publishing, 117–211.

Wolfe, Alan. (1989). *Whose Keeper?: Social Science and Moral Obligation.* Berkeley: University of California Press.

Yurow, Jane H., with Robert F. Aldrich, Robert R. Belair, Yale M. Braunstein, David Y. Peyton, Stanley Pogrow, Lawrence S. Robertson, and Aaron B. Wildavsky; edited by Helen A. Shaw (1981). *Issues in Information Policy.* Washington, DC: GPO.

AUTHOR INDEX

183

SUBJECT INDEX

www.ingramcontent.com/pod-product-compliance
Lightning Source LLC
Chambersburg PA
CBHW050439280326
41932CB00013BA/2172